Semantic Control for the Cybersecurity Domain

This book presents the creation of a bilingual thesaurus (Italian and English), and its conversion into an ontology system, oriented to the Cybersecurity field of knowledge term management and the identification of a replicable method over other specialized areas of study, through computational linguistics procedures, to a statistical and qualitative measurement of the terminological coverage threshold a controlled vocabulary is able to guarantee with respect to the semantic richness proper to the domain under investigation. The volume empowers readers to compile and study significant corpora documentations to support the text mining tasks and to establish a representativeness evaluation of the information retrieved. Through a description of several techniques belonging to the field of linguistics and knowledge engineering, this monograph provides a methodological account on how to enhance and update semantic monitoring tools reflecting a specialized lexicon as that of Cybersecurity to grant a reference semantic structure for domain-sector text classification tasks.

This volume is a valuable reference to scholars of corpus-based studies, terminology, ICT, documentation and librarianship studies, text processing research, and distributional semantics area of interest as well as for professionals involved in Cybersecurity organizations.

Semantic Control for the Cybersecurity Domain

Investigation on the Representativeness of a Domain-Specific Terminology Referring to Lexical Variation

Claudia Lanza

CRC Press
Taylor & Francis Group
Boca Raton London New York

CRC Press is an imprint of the
Taylor & Francis Group, an **informa** business

First edition published 2023
by CRC Press
6000 Broken Sound Parkway NW, Suite 300, Boca Raton, FL 33487-2742

and by CRC Press
4 Park Square, Milton Park, Abingdon, Oxon, OX14 4RN

CRC Press is an imprint of Taylor & Francis Group, LLC

ISBN: 978-1-032-25080-9 (hbk)
ISBN: 978-1-032-25081-6 (pbk)
ISBN: 978-1-003-28145-0 (ebk)

DOI: 10.1201/9781003281450

Typeset in CMR10
by KnowledgeWorks Global Ltd.

Publisher's note: This book has been prepared from camera-ready copy provided by the authors.

Contents

List of Figures ix

List of Tables xi

Foreword xiii

Preface xv

1 Background 1
   1.1 Corpus linguistics 1
      1.1.1 Corpus representativeness 1
      1.1.2 Thesauri term representativeness 5
      1.1.3 Corpus design 10
   1.2 Text mining 12
      1.2.1 Data pre-processing 13
   1.3 Terminological knowledge bases (TBK) 15
      1.3.1 Automatic term extraction (ATE) 17
      1.3.2 Gold standards reference 19
      1.3.3 Domain-dependency 20
      1.3.4 Terms population 20

2 Case study: Cybersecurity domain 23
   2.1 OCS project 23
   2.2 Specialized languages reflecting specialized domains 25
   2.3 Cybersecurity field of knowledge 27
   2.4 Corpus design for Cybersecurity domain in Italian language . . 29
   2.5 Existing resources for Cybersecurity domain 31
      2.5.1 Italian resources 32

		2.5.2	English resources	34
	2.6		Group of experts supervision	35
3	**Related works**			**37**
	3.1	KOS		37
		3.1.1	Thesauri	39
		3.1.2	Ontologies	45
	3.2		Semantic conversions – from thesauri to ontologies	48
	3.3		SKOSs systems	50
	3.4		Research approaches	50
		3.4.1	Clustering approaches	51
		3.4.2	Keyphrases extraction	52
		3.4.3	NLP approaches for building semantic structures	54
			3.4.3.1 Distributional similarity	54
4	**Research methodology**			**61**
	4.1		Corpus construction	61
		4.1.1	Documents selection	62
		4.1.2	Software-aided terminological extractions – First selection	69
			4.1.2.1 T2K – Linguistic-oriented tool	71
			4.1.2.2 TermSuite	74
			4.1.2.3 Pke-keyphrases detection	79
			4.1.2.4 Results	82
	4.2		Candidate terms selection	83
		4.2.1	Frequency criterion	83
		4.2.2	Mapping with gold standards	85
		4.2.3	Expert support system	90
	4.3		Automatization of thesaurus construction	93
		4.3.1	Approaches	93
		4.3.2	Variants	94
		4.3.3	Patterns-based	94
			4.3.3.1 Causative relations	95
			4.3.3.2 Hierarchy	98
			4.3.3.3 Synonyms	98
		4.3.4	Word embeddings detection	99
5	**Semantic tools for Cybersecurity**			**105**
	5.1		Construction of Italian Cybersecurity thesaurus	105
		5.1.1	Semantic relationships	108
			5.1.1.1 Hierarchy	108
			5.1.1.2 Equivalence	109
			5.1.1.3 Association	110

5.1.1.4   Scope notes 110
5.2   Ontology conversion 112
5.2.1   Structure 115
5.3   Discussion 123

6   Semantic enhancement and new perspectives **127**
6.1   Multilingual alignment 127
6.2   Monitoring of terminological representativeness 129

7   Conclusion **135**

References **147**

Index **165**

List of Figures

1   PhD research phases xxi

1.1 Approaches for corpus representativeness. 4

2.1 Statistics OCS website 25
2.2 MeSH tree view for virus 27
2.3 Hierarchy of laws 30
2.4 Cybersercurity National structure as reported in the White Book on Cybersecurity (2018:18) [147] 32

3.1 KOS diagram classification [127] 38

4.1 Magazines from Ulrich's web and BNCF for the Cybesecurity domain 64
4.2 Results of termhood measures 85
4.3 Synthesis of the main first candidate terms with reference to the shared knowledge contained in the evaluation lists 89
4.4 New terms derived by the collaboration with the group of experts 91
4.5 Comparison of RT in thesaurus and patterns paths outputs 96

5.1 Subject category and broader term 108
5.2 Cybersecurity hierarchical structure 109
5.3 Equivalent terms 110
5.4 Related terms 111
5.5 Scope Notes for terms in the thesaurus 113
5.6 Ontology graph for the Italian Cybersecurity thesaurus conversion 116
5.7 Thesaurus representation of the semantic relationship that describes opposition 118

5.8   Protégé representation of the semantic relationship that describes
       opposition 119
5.9   Ontology representation for vulnerability connections 119
5.10   Ontology representation for actors: Hacking with hackers and
       crackers 119
5.11   Ontology representation of antispyware and malware
       connection 120
5.12   Additional ontology Object Properties through patterns
       path-variables 120
5.13   Ontology representation of Security properties as *Data Proper-
       ties* 120
5.14   Thesaurus representation of Security properties as hierarchical
       relations 121
5.15   Crackers' individuals 122
5.16   Individuals for malicious software representation 122
5.17   Associations retrieval in ontology by patterns configurations . . 123

6.1   RDF node 130

List of Tables

4.1   Corpus criteria selection 65

4.2   Italian Cybersecurity corpus size 68

4.3   Types Tokens Hapaxes in the source corpus 69

4.4   Term extraction details 83

4.5   Reference quantitative population of main multilingual Cybersecurity gold standards 86

4.6   Terminology mapping results source corpus with the gold standards NIST, ISO, WhiteBook, and Glossary of Intelligence (GLOSS) 86

4.7   T2K, TermSuite, PKE, and BERT term extractions executed on Cybersecurity corpus. The evaluation is performed over five lists (Clusit, Glossary, Nist, Iso, and Cyber) and the results (%) are provided with respect to Precision (P), Recall (R), and F-measure (F1) 90

4.8   Output of the semantic connections extractions run with word2vec (W2V) and fastText models using the Precision at 100 (P@100%) score. 99

5.1   Cybersecurity ontology metrics on Protégé platform 117

Foreword

This book presents the creation of a bilingual thesaurus (Italian and English), and its conversion into an ontology system, oriented to the Cybersecurity field of knowledge term management and the identification of a replicable method over other specialized areas of study, through computational linguistics procedures, that can measure from a statistical and qualitative perspective the terminological coverage threshold a controlled vocabulary is able to guarantee with respect to the semantic richness proper to the domain under investigation. The volume empowers readers to compile and study significant corpora documentations to support the text mining tasks and to establish a representativeness evaluation of the information retrieved. Through a description of several techniques belonging to the field of linguistics and knowledge engineering, this monograph provides a methodological account on how to enhance and update semantic monitoring tools reflecting a specialized lexicon as that of Cybersecurity to grant a reference semantic structure for domain-sector text classification tasks.

This volume is a valuable reference to scholars of corpus-based studies, terminology, ICT, documentation and librarianship studies, text processing research, and distributional semantics area of interest, as well as for professionals involved in Cybersecurity organizations.

Preface

Problem statement

The purpose of this study is that of developing a semantic tool, a thesaurus – which will be further converted to an ontology structure – to systematize the information about Cybersecurity in Italian language and to provide a methodology to measure the representativeness of the semantic resource reproducing the terminology of the specialized domain alongside the semantic structure derived from a computational analysis of the terminology within the field of study. The project aims at creating a semantic monitoring tool for the Cybersecurity domain, which currently has not been officially provided, that can help organizations to better frame the information on Cybersecurity and provide a terminological means of support that could be useful to broadly understand the domain from a semantic point of view. Indeed, the research project is inserted within the Cybersecurity specialized field of knowledge, and the terminology included in its set of data, that makes up the source corpus exploited to process the information through NLP tasks so to realize a semantic structure that gathers the information related to it, reflects the technical trait of the domain [67]. The project described takes its ground on the objective of developing a semantic structure in Italian language for Cybersecurity, which can be considered a reliable source to guide users in the comprehension of specific semantic nodes of the domain as well as a means to study the ways by which the terminology base is created to become the basis from which variation through time will be stored. Although there are plenty of books or guidelines on the domain of Cybersecurity, such as, for example, the Glossary edited by the Italian Presidency of Ministers "Glossary of Intelligence", the "White Book of Cybersecurity" published by the Italian University Informatics Consortium CINI group [147], or the guidelines published by the Computer Emergency Response Team (CERT), this study proposes the thesaurus as the semantic tool to structure its terminology since it can be considered a tool that could give a more detailed overview of specialized domains of study thanks to the possibility of creating relationships between the terms that are meant to be representative of specific fields of knowledge [171]. Moreover, given that one of the phases pursued in the project activity regarded the creation of the Cybersecurity ontology, the semantic tangle that comes out by the creation of a thesaurus is a starting point that facilitates the process of migrating the knowledge organization within it into an ontology system [54] and, consequently, guaranteeing a higher interoperability between systems to process domain-oriented data. The main intentions of this research project are:

- Create an Italian thesaurus on Cybersecurity, currently not existing, that can help organizations to better frame the information on Cybersecurity and provide a terminological means of support meant to help the comprehension of the technical type of information proper to this specific area of study;

- Set up a methodology that could be applicable to multiple sector-oriented contexts that can evaluate the optimal threshold level for a semantic resource that can establish if it represents a technical domain, that means if it is or is not adequately and semantically covering the specific area of study knowledge.

In particular, this scientific investigation analyzes the multiple sides related to the terminological knowledge base enhancement in retrieving the most representative candidate terms of Cybersecurity domain in Italian language starting with the main gold standard repositories and authoritative sources, going further towards a progressive enrichment from the social media context, which gives more information about the latest trends on terms' usage within the specialized world to be analyzed and represented through semantic resources such as thesauri and ontologies. To achieve these previous results, the following chapters focus on the methodologies and software employed to configure a system of NLP rules to get the desired semantic outputs and to proceed with the enhancement of the candidate terms selection which are meant to be inserted in the controlled vocabulary. Given that the semantic resources realized to manage the Cybersecurity terminology extent are two, i.e., a thesaurus and an ontology created under the basis of the thesaurus structure, not only a system of programming rules that has helped the configuration of the first semantic levels of connections between terms in the thesaurus will be provided, but also a discussion on how an ontology can better systematize the knowledge of this technical domain and improve the interoperability among systems by using OWL and Resource Description Framework (RDF) language will be presented.

Thesaurus and Ontology

The reason why these two semantic means of knowledge organization and representation have been selected is based on the assumption that they can provide a highly structured way of managing a specialized domain's information through precise semantic networks. Moreover, both of them can be realized after a computer-assisted manipulation of the texts that allows both to treat natural languages as structured chains of textual data to be integrated with frequency scores, and to apply on the post-processed text distributional semantics methodologies to discover the semantic relationships among terms that will configure the structure of these semantic resources. In particular, a thesaurus is a "controlled and structured vocabulary in which concepts are represented by terms, organized so that relationships between concepts are made explicit, and preferred terms are accompanied by lead-in entries for synonyms or quasi-synonyms" [83]. Still in the ISO 25964:2011, it is possible to find its main objective, that is to "guide both the indexer and the searcher to select the same preferred term or combination of preferred terms to represent a given subject. For this reason a thesaurus is optimized for human navigability and terminological coverage of a domain." Therefore, a thesaurus can guarantee a domain-oriented systematization of its specialized information by managing the relations within its concepts, which are in turn represented by terms, in a fine-grained network of semantic connections, which can help to semantically understand the domain. In organizing the representative terms in a hierarchical way, as well as under the equivalence and association perspective, the terminologists should comply with standard guidelines in order to realize a tool that will provide a normalized information systematization accepted by every technical sector for practical applications. This last feature proves to become quite relevant when experts of the domain or areas of study interested in carrying out some activities in the field of knowledge semantically organized by a thesaurus may need a standardized list of terms, relationships, and definitions on which they can rely to cohesively work in the targeted sector [95]. The thesaurus itself represents a reliable means of semantic control that can orientate the overall understanding of specialized domains. The intrinsic nature of the thesaurus is to help indexers in the process of selecting the preferred terms and connecting them through a series of pre-established relationships, but the way by which it can be looked into is limited by its semantic relationship fixity. For this reason, and as it will be deeply discussed in the chapter dealing with the ontology migration (section 5.2), the thesaurus has also been converted into another semantic tool to represent a specialized field of knowledge, i.e., an ontology. The benefit that comes from the usage of ontologies is related both to the languages (OWL and RDF) in which they are written that allow for a higher form of interoperability between informative systems, a feature that can be very helpful in sharing specific information within a broad community of the domain's experts, and to its more flexible nature in representing the information proper to technical domains. Being a means usually employed to structure specialized fields of knowledge, ontologies are taken into account among several areas, e.g., NLP, knowledge representation, knowledge management, web services, etc. Ontologies, as stated by Gruber (1993:1):

> 'An ontology is an explicit specification of a conceptualization. The term is borrowed from philosophy, where an Ontology is a systematic account of Existence. For AI systems, what "exists" is that which can be represented. When the knowledge of a domain is represented in a declarative formalism, the set of objects that can be represented is called the universe of discourse. This set of objects, and the describable relationships among them, are reflected in the representational vocabulary with which a knowledge-based program represents knowledge. Thus, in the context of AI, we can describe the ontology of a program by defining a set of representational terms. In such an ontology, definitions associate the names of entities in the universe of discourse (e.g., classes, relations, functions, or other objects) with human-readable text describing what the names mean, and formal axioms that constrain the interpretation and well-formed use of these terms. Formally, an ontology is the statement of a logical theory.'

These resources are characterized by a higher level of conceptual abstraction, and also by different restriction rules that can better describe technical knowledge through the usage of formalisms, such as cardinality or axioms. The realization of an ontology for the Cybersecurity domain is linked to its more effective way of representing associative relationships existing among the preferred and non-preferred terms. As it will be shown in the following sections of this book, the ontology has been preferred over the thesaurus, or better joined to the thesaurus structure, because of the latter's flat visualization of several relations among terms. Specifically, the ontology enables users to assign explicit forms of relationships between classes, the so-called *Object Properties* and *Data Properties*, which represent the predicate connecting two classes and the attributes to assign. For example, *cyber war* RT *cyber weapon*, is made more explicit in the ontology under the following form: *cyber war* uses *cyber weapon*, where 'used' represents the ObjectProperty.

Objectives and Scope

The goal of this research project is, therefore, to provide solid semantic tools that can gather the technical information on the Cybersecurity domain with the objective of achieving a terminological coverage as complete as possible. The scope of this study is represented by the Cybersecurity Observatory project designed for small-medium enterprises and public users to provide a web-based platform to guide the cyber defence strategies, provide alerts on new malware or spam campaigns, and to orientate the understanding of the technical lexicon proper to this domain. To reach this latter perspective, the project aims at enhancing the terminological set of data by taking into account not only documents coming from the legislative world and from sector-oriented magazines, i.e., the main starting information channels to build on the source corpus needed for the terminological extraction process, but also social media infrastructures and, doing so, achieve a heterogeneous information repository that empowers the data strength about the domain. In order to achieve these objectives, the collaboration with experts proved to be fundamental. Currently this semantic resource is present among the services on the website developed by the CybersecurityLab in Pisa (Italy). This online platform has been created for experts and common users to orientate their comprehension of the Cybersecurity domain. The thesaurus contains 245 terms, almost all of them with their definition extracted from the corpus documents and with their equivalents in English. Their integration has been established by executing different terminological extractions on authoritative and popular sources and then by applying pattern configurations, variation techniques, and word embedding algorithms to automatize the relationships' connections detection.

The results that have been preeseded are linked, in the first place, (i) to the realization of the thesaurus and the ontology by pursuing an information retrieval technique on authoritative documents on the domain, then (ii) to statistical measurements related to the Type Token Ratio (TTR) to underline the lexical richness and variance inside the corpus from which to start to select the candidate terms, and (iii) through Precision and Recall measures based on the output of the terminological extractions with the main gold standards existing on the Cybersecurity field of knowledge; afterwards, (iv) the mapping system between the flat list of the thesaurus with the gold standard by using Python NLP language libraries proved the level of accuracy the semantic tools have been able to reach with respect to the terminology officially shared by the experts; the automatic retrieval of the semantic relationships starting from the word embedding algorithms (v) and pattern configuration inside the source corpus, with specific connection to the causative rule-based pattern code, and through variation detection using semi-automatic terminological extraction software (vi); as final results there are the implementation of SPARQL (vii) query system on the ontology to make the semantic platform readable from several informative systems and let it become a searchable database to be integrated with the latest legislative and popular documents published by reliable sources in order to automatically populate the conceptual framework; and the (viii) new social media terminology on the domain detection, especially referring to neologisms on new arising malware, this should be a trigger that can indicate a change in the information tissue of the domain and the ability to capture the variations that could increase the proximity of the digital semantic tool for the Cybersecurity representative with respect to the domain's terms evolution.

Structure of the book

This book starts with a presentation of related works for what concerns the general frameworks of corpus linguistics as well as the processing of terminological extraction and references to gold standards repositories. Chapter 2 is centred on the applied Knowledge Organization Systems (KOSs), thesauri, and ontologies; Chapter 3 presents the case study of Cybersecurity by giving details on the context and highlighting the support of the group of domain experts. Chapter 4 is specifically based on the description of the methodology followed, while Chapter 5 focuses on the phase that refers to the semantic resources' enhancement. Chapter 6 includes some future works to carry out related to the project scope, the SPARQL query system and tweets profiling. The work ends with the Conclusion section that presents a synthesis of the main points described in the previous parts of the book. In the appendices, the structure of the thesaurus and ontology configurations will be described with a few examples to clarify the main differences occurring between the two semantic systems in the way of managing the specialized knowledge domain for the Cybersecurity framework.

Project Phases

For a better understanding of the previous passages, the following workflow depicts the steps followed for the thesaurus and the ontology construction.

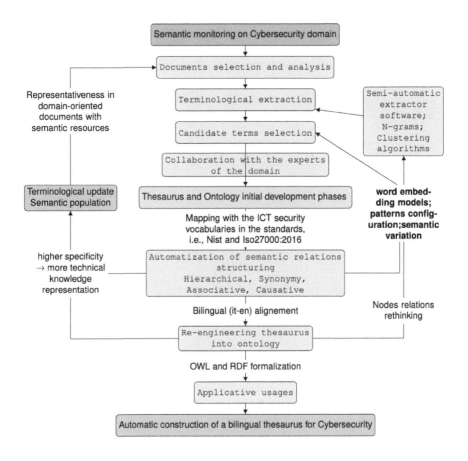

Figure 1: PhD research phases

Chapter 1

Background

1.1 Corpus linguistics

This chapter first describes the main theories that have been shared by experts in linguistics, knowledge engineering, Natural Language Processing (NLP), and information retrieval groups, referring to the measurement of corpora representativeness given their domain-oriented nature, and properly to the different perspectives in the literature for the corpora composition or design. Successively, this preliminary section focuses on the description of the processing phases of the information, collected within the specific-domain corpora, highlighting the key approaches involved in text mining procedures, e.g., the data pre-processing operations, the grouping systematization of texts or their categorization. Finally, this part presents some of the general features dealing with the construction of semantic resources, in particular with reference to the gold standards matching systems, the issue of domain-dependency of the terminological bases and how the lists of terms which are representative of technical domains can be increased by semi-automatic population models.

1.1.1 Corpus representativeness

A corpus is:

> "A collection of pieces of language that are selected and ordered according to explicit linguistic criteria in order to be used as a sample of the language. Words such as 'collection' and 'archive' refer to sets of texts that do not need to be selected, or do not need to be ordered, or the selection and/or ordering do not need to be on linguistic criteria. They are

DOI: 10.1201/9781003281450-1

therefore quite unlike corpora. Linguistic criteria to be applied to the selection and ordering may be: External
– in that they concern the participants, the occasion, the social setting or
the communicative function of the pieces of language; Internal
– in that they concern the recurrence of language patterns within the pieces
of language." (EAGLES 1996e) [1]

As stated by Sinclair (1995), the representativeness of a corpus should be
evaluated by considering external factors, meaning that the documents intended
to be part of the corpus should be selected taking into consideration external
criteria related to population targets from a linguistic and communicative perspective in a progressive way. Following this principle, Biber, Clear or Osterl
and several other linguistics experts in corpus design shared the idea that the
coverage of corpora can be related to the application context to which the semantic resources will be effectively applied. Therefore, the domain specificity of
language should be considered as one of the main features to take into account
to calculate the semantic tools coverage threshold, and variation of terminology
over time. Hunston (2002) [155], for instance, suggested the assumption of considering the corpus either as a *static* or *dynamic* reference. In the first case we can
refer to a corpus as a sample of documents that may represent a domain, in the
second one the corpus compilation refers to the monitoring of semantic chang-
ing. As clarified across the works published by the Lancaster University section
for studying corpora[1], *static* models can present several opportunities to support
a very efficient analysis on a cross-internal language variation in the documents
exploited, transforming them into *sampling frames*, which, in turn, are crucial
means to condense the scope of a specialized corpus. In this latter work the author provides a further explanation of the levels of representativeness granted by
corpora by referring to a specific or general corpus, underlying the different ways
in which the calculation of their representativeness is getting measured. Indeed,
even if for both of them the level of semantic density is given by an inclusion
of as many texts as possible meant to represent the variety of a lexicon, they
do differ in their measures. Following the suggestion of the author, *generalized*
corpora tend to compute their representativeness based on the incorporation of
several types of texts, whereas for what concerns the *specialized* ones, the focus
should be kept on the range of "closure" or "saturation" of the corpus (2002:3).
The concept of saturation is followed in this book. It considers the moment in
which the tokens proper to the texts in the source corpora will not be expanded
in the deeper analysis of a lexicon variation over time. At the point in which,
from a lexical perspective, the key terminological elements of a specialized corpus, which is meant to be a reflection of technical domains of study, reach their
saturation, meaning that any other new entries in the lexical vocabulary would
not provide any crucial semantic increase and an added value in the terminology

---

[1] http://corpora.lancs.ac.uk/clmtp/answers.php?chapter=1&type=questions, Accessed 15/04/2020

system represented, the corpus should be considered as covering the information inside a given area of study. This is possible, for example, by designing a corpus of documents in which the information is contained under different formats. The genres of texts are quite significant in the way they create a balancing system for reference corpora. What makes a corpus balanced is the semantic granularity of the information contained in several categories of texts. It is quite difficult, as pointed out by Hunston (2002), to catch up with continuing changes over time in the documents that constitute the reference corpora. To reach a semantic coverage, according to this latter theory, should be important to keep in line with the terminological variation in the specialized fields of information, getting closer to the *dynamic* corpora models. These latter seem difficult to manage in order to keep a good distribution of novelties arisen in a technical lexicon to be analyzed. It is for this reason that recent literature on corpus representativeness specifies the importance of "size" in evaluating the quality a corpus holds [64], even referring to its accuracy in collecting the right high-technical information of a specialized domain. Many experts in corpus linguistics (Sinclair, Leech, Condamines, and Biber) shared the idea that the value of a corpus can be attributed to its numerical proportion of the texts in a given moment.

> "The proportion and number of samples for each text category. The numbers of samples across text categories should be proportional to their frequencies and/or weights in the target population in order for the resulting corpus to be considered as representative." (Indurkhya, 2010:153) [157]

Even though the total inclusion of the linguistic units in a corpus is very hard to finalize, it is still crucial having in mind the communicative intention with which the corpus has to be designed, its scope and audience, as well as its stratification of sampling population to reach. Referring to works of Rapp (2014) [151] and Francom (2010) [85], representativeness could also be given by the frequency of words in a corpus, their co-occurrence and the common words in the texts.

> "A corpus whose word frequencies are highly correlated to the familiarity norms is more likely to be a good surrogate for everyday language." (Rapp, 2014:2118)

Indeed, the study carried out by Francom (2010) addressed to the description of how corpus representativeness can be gauged by frequency effects. The familiarity of words given a number of semantic units related with their frequencies calculated from a corpus processing is one of the key indicators that may determine to which extent a corpus can represent the variability of a lexicon shared by a target population.

According to Gries (2010) [159], and as this study will focus on, the coverage a domain-specific corpus retains can be achieved by referring to an external one validated by experts source. Representativeness' computation has been dealt

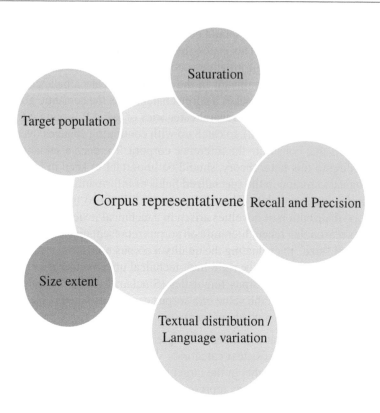

Figure 1.1: Approaches for corpus representativeness.

as a key element to corpora appraisal by many authors in corpus linguistics, they proposed several approaches, as, for instance, Bel (2004) [136] who indicates the strategy for measuring the appropriateness of domain languages used in specific fields of knowledge. The author explores the semantic quality inside the texts included in a domain-specific corpus through the employment of NLP techniques to induce syntactic accurate information proper to specific areas of study as indicators of corpus domain coverage. When it comes to describing the corpus coverage, Bel uses the word "full coverage" to address the status of the typologies of words used in a specialized domain and the semantic information needed to be parsed to achieve every word occurrences. The author underlines the importance of having a set of data specifically and uniquely related to a very technical field of knowledge, as the case of study of this study hereby proposed, which holds its proper grammatical and lexical features markedly characterizing it from others domains. Using the words of Gray (2017) [34], the domain-specificity issue can be related to what he calls the *target domain representativeness*, which means the extent to which a corpus assembles specific domain texts to fully cover its definite semantic variability. Whereas, the *linguistic representativeness* is the complex of all the linguistic distributions in a corpus. Another work that refers to the way corpora can be compared to the variability of the information included in order to achieve the right proportion of informative distribution is that of Gablasova (2019) [47].

The extent of the distribution of certain domain-specific words in a set of documents that compose a source corpus is a key element to recognize their "termhood" level. As Kageura (1996) [97] stated, and then resumed by Hisamitsu (2000) [78] in his work on the discovery of a method to measure terms representativeness by considering their frequency in a corpus, the termhood is the degree to which the lexical units are related to domain-specific concepts. Even though, as Nakagawa (2002) pointed out, termhood is calculated by applying statistical measurements of frequency or inverse document frequency, and this implies that sometimes terms own specialized meanings according to their application usages and, perhaps, not relating with writer's intentions [76]. This can widely open the perspectives for a fine-grained way to conceive the scope of semantic representativeness in a corpus.

Among the statistics studies on the coverage a set of documents can show with respect to a highly specialized domain of study, the *Precision* and *Recall* formula is commonly exploited by most of the experts in this discipline. This calculation refers to the accuracy check of a system measuring the terminological exhaustiveness and precision inside a controlled dataset in order to guarantee a retrieval of all the relevant terms related to a research starting from a particular database. This calculation statistically provides the percentage of relevant and irrelevant records in the researches: recall is the measure of the numbers of relevant records retrieved from the total number of relevant terms in the database; precision, on the other hand, is the measure of the total of relevant terms retrieved compared to the total number of retrieved terms. Even though this measurement appears to be an important evaluation when it comes to considering the coverage a thesaurus can reach with reference to a specialized corpus terminology, it does not give a satisfactory criterion for a wide-ranging measurement of the representativeness of the semantic tool related to geographic contexts since it takes into consideration a pre-established set of documents to verify the terminological coverage and the precision in the information retrieval [173].

1.1.2 *Thesauri term representativeness*

Many studies on evaluating theasauri quality effectiveness have explored ways in which these semantic resources could be considered as reusable systems to guide towards technical domain understanding [146]. In the literature there are some examples of the approaches undertaken by experts in applying a set of statistical measures able to discover the reliability of a semantic tool like a thesaurus. For instance, Virginia (2011) [70], proposed a methodology to compute the thesaurus effectiveness using measurements commonly exploited in information retrieval, i.e., Recall and Precision, and going further by generating a tolerance matrix with lower and upper sets, and then creating a tolerance rough set, which has been considered like an associative thesaurus. The Tolerance Rough Set Model (TRSM) that the authors took as the starting point represents a measure typically employed for data classification and, as the authors stated:

> "The central point of rough set theory is based on the fact that any concept (a subset of a given universe) can be approximated by its *lower* and *upper approximation*." (2011:706)

Analyzing the details of Virginia's (2011) work, at each level, the approximation space will be a set of documents represented by a weight vector in which a *tolerance class* of terms, conceived as a positive parameter, is associated to documents. Through the variation of this tolerance class, which the authors define as "threshold θ,"

> "[...] one can tune the preciseness of the concept represented by a tolerance class." (2011:707)

This procedure has been tested on two specialized corpora to discover the preciseness the concepts represented by terms, retrieved after a previous process of text pre-treatment, e.g., stemming and tokenization, being part of a set of vectorized documents. They took a thesaurus as a "term-by-term matrix which contains tolerance classes of all index terms" (2011:708). The accuracy of these terms exhibits, measured also by the Recall and Precision metrics, reveals the specificity of concepts that constitutes the source set, being thus a discriminating value to ensure the coverage of a semantic tool with respect to precise document sets. In this specific use case, the Recall and Precision are calculated on terms rather than documents, so the Recall represents the proportion of relevant retrieved terms, and the Precision the proportion of those that are relevant. Hence, for the authors, the preferred higher value between Recall and Precision is the first one since it guarantees the presence of significant semantic units, i.e., terms in the source set. It is possible to deduct that the higher percentage of relevant terms which have been gathered to develop a semantic resource can be the reflection of specialized domains of study. This measurement line is pursued across this book as scale comparison for thesaurus representativeness.

On the same principle of Virginia, other studies have been focused on the analysis of thesauri reliability and on a set of disparate approaches to evaluate the quality of these resources from a re-usability perspective and a domain coverage scope. Mader (2012) [43] gave, for instance, 15 quality issues for SKOS vocabularies, which are commuted into quality functions working as assessment categories to check the validity of semantic tools. Among these decision-making quality functions, tested on several SKOS frameworks as matching contexts (e.g., *MeSH, AGROVOC, DBpedia, Eurovoc*), there are, for example, the *orphan terms, cyclic hierarchical relations, incomplete language coverage and broken links.* These categories underline certain circumstances under which sets of terms composing the source corpora exhibit and can reveal a starting point to be adjusted in order to have, in a future perspective, a highly accurate semantic resource and not have problems in bringing it together with other ones in terms of interoperability. Indeed, as stated by the authors, if in a reference semantic tool there is a range of broken links, this, perhaps, could hinder the interconnection with other resources that may create a networked mechanism for sharing technical knowledge.

A parameter that can support the evaluation phase to verify the accuracy a semantic tool, like a thesaurus, holds, is the availability of expertise in the domain of study. This feature is often provided by the figure of the group of experts of the domains [161] supporting the validation steps toward the candidate terms inclusion. By gaining the consensus of domain experts, the activity of the terminologists in making the semantic resources represent a particular domain can be facilitated since it is overseen by people working day-by-day in the targeted context. Indeed, the co-operation with these figures is a crucial step to check the validity of the structured terms representing the concepts of specialized fields of knowledge. The jointed collaboration between experts and terminologists typically allows to create reliable semantic resources made of the most precise and commonly used terminology of the domain to be represented to guide its understanding. In fact, one of the main objectives addressed by the terminology is the maintenance of the updated way experts communicate with each other and formalize this interconnection through standardized semantic methodologies able to develop guided controlled vocabularies or ontologies gathering technical information to be re-used in application contexts. Whichever will be the semantic resource terminologists will choose alongside the specific needs of several organizations, the key point is to provide a means through which users can identify the domain's updated information framework. As a result, the tool generated as a response to these requirements, should be a reflection of the current conceptual model of a domain, thus it should be as much updated as possible guaranteeing the semantic coverage, as this study aims at achieving for the Cybersecurity field of knowledge.

A statistical method to measure the extent a corpus should present, and, consequently, how this score may influence the reliability of a semantic tool generated to structure the information contained in a source corpus, is proposed by Trunfio *et al.* (2014) [10]. The authors provided an approach to calculate if a sample size of a corpus is sufficiently extensive or has to be enhanced with other texts by applying a *Rinott Procedure* statistical measure [179]. Their methodology is based on taking into consideration an initial sample size, which they called n_0 composed of the corpus' texts, carefully selected *a priori* according to quality guidelines and time-based assumptions, then:

> "[...] for each text t in the sample corpus, with $t = 1, \ldots, n_0$, let I_t be an index of the lexical richness of text t. Observe that all the measured indices are independent under the assumption that the texts are suitably

selected. Successively, measure the unbiased estimate of the sample variance $S_{n_0-1}^2 = \sum_{t=1}^{n_0}(I_t - \bar{I})^2$ where

$$\bar{I} = \sum_{t=1}^{n_0} I_t / 2$$

is the sample mean of the selected index. Thus, the minimum corpus size with the desired error and confidence level can be measured as follows (2014:3):

$$n = max\left\{ n_0 \left\lceil \left(\frac{z_{\alpha/2} Q \cdot S_{n0-1}}{\varepsilon} \right)^2 \right\rceil \right\}"$$

Therefore, if the sample size n is greater than n_0, the authors consider to increase the proportion of the documents that have been inserted in the sample corpus, which they specifically refer to as $n - n_0$.

One of the measures Trunfio *et al.* addressed to in their work is a commonly exploited means to compute the lexical richness and vocabulary diversity in given corpora, i.e., *Type-Token-Ratio* [99]. This calculation given by dividing the number of tokens, that constitute the total number of words in a given set of documents, by the number of types, which, by contrast, represent the different words occurring in given texts, tells the percentage of repetition in a source vocabulary. The higher the number of the score, the more varied the vocabulary is within the source texts [37].This measure, together with the proportion of *hapaxes* in the vocabulary [131], which represent words that occur very infrequently and just once, may highlight how specific a list of candidate terms to be included in a semantic resource like a thesaurus or an ontology can be evaluated, and for this reason it is meant to be employed to show the results for the Cybersecurity domain terminology.

Kaguera (2018) [150] also dealt with this potential estimation of size coverage for corpora, and proposed a method by which he extracts the size of terms occurrences observed in a given corpus of Japanese and English texts on municipal domain, mostly analyzing compound nouns, *to infinity* (2018:87) and evaluating the *saturation point*. As a second step, the author studies the up-to-date status of the range of types by comparing it with the saturation point of the first phase. Moreover, he points out that to evaluate to which extent the terminology for a given domain is well formed, the set of conditions to take into account are the controlled ones, as they can grant a greater stability in the variation of languages, since terminologists may select the preferred structures to be exploited, and facilitate the fixity of certain standardized linguistic expressions to be studied. The variation of terms in given vocabularies [29] is a remarkably important feature to consider to develop semantic tools meant to represent in a standardized form a specific field of knowledge information. To be able to run up with the most recent updates in the specialized domains terminology to be represented, terminologists' activity should be focused on detecting and managing the current semantic changes. As a consequence, the result should be the enhancement of the terms systematization that normally guides the understanding of the domain information. For this reason, the attempts to achieve terminological representativeness have as a perspective that of collecting the most used standardized form of representative terms and their accepted variants by a community of experts or by the authoritative sources being the main channels from which terms come from. Apart from the aforementioned statistical approaches, and other techniques that will be better described in the next sections dealing with corpora composition rules, this book is also focused on the study of computational methodologies that identify the current terminology updates. The main objective is to capture the most recent changes to be able to provide reliable updated semantic resources by monitoring the main networks where the source documents come from and retrieving other elements in social media. The second result is to create a system that can facilitate the detection of new terms in the Cybersecurity domain related to several semantic families, as for example those of *cyber threats*, or *cyber crime*. These procedures are meant to be started from the theoretical principles of the *business intelligence*, or *veille informatique*. The main purpose of this detector methodology is that of capturing the latest updates in the field of study with which business users interact [75], in this way, through a set of pre-selected sources of information which share targeted contents, the technical ground can be constantly renewed, and, as a result, more trustworthy.

This way of getting updated with the recent changes in the contents published by a range of pre-established sources may guarantee an upgraded version, and, by consequence, more reliable, of a semantic tool if connected with these flows of new information.

Another way to verify the accuracy and preciseness of a corpus proper to a specialized area of study could be the comparison with existing certified bodies of knowledge, i.e., gold standards [138], as described in section 1.3.2. This study aims at reaching a sufficient threshold of coverage of a specialized domain, i.e., that of Cybersecurity, through (i) the mapping procedure with the gold standards of the domain to be studied, i.e., the matching procedure in aligning representative terms chosen by the extracted ones with the semantic units officially accepted by groups of experts [161]; (ii) the variation detection of the terms lists given by the terminological extractors tools and methodologies, including the definition of the semantic relationships to define a supervised semantic structure of a field of knowledge to avoid repetitions and obsolescence issues [30]; (iii) the observation of new incoming lexical representative units given by the updated contents of the sources of information making up the reference corpus and social media streaming word associations. These points will be applied both on the thesaurus and the ontology on the Cybersecurity domain. In particular, the monitoring phase, which will infer the parameters according to which the semantic tool should or should not be updated, will be especially executed on the ontology given its greater flexibility and the possibility to work with more variables in *SPARQL* to discover novelties in the reference sources. Following the recommendations provided by the W3C Consortium, SPARQL queries usually consist of a:

> "[...] set of triple patterns called a basic graph pattern. Triple patterns are like RDF triples except that each of the subject, predicate and object may be a variable. A basic graph pattern matches a subgraph of the RDF data when RDF terms from that subgraph may be substituted for the variables and the result is RDF graph equivalent to the subgraph. [...] The query consists of two parts: the SELECT clause identifies the variables to appear in the query results, and the WHERE clause provides the basic graph pattern to match against the data graph. The basic graph pattern in this example consists of a single triple pattern with a single variable (?title) in the object position."

1.1.3   Corpus design

As specified by Bowker (2002) [106], "A corpus is not only a random collection of texts [...]. Rather, the texts in a corpus are selected according to explicit criteria in order to be used as representative sample of a particular language or subset of that language."

In order to have a set of documents that should characterize the source corpus, and by consequence creating a semantic tool, such as a thesaurus or an ontology, the process of choosing specific parameters in filtering the informative tissue proves to be quite an important step. Corpus linguistics [87], according to specific research requirements, suggests to take into consideration documents that are related to present age, better if they share the same time range and are all written in the same language, including produced in a same national area. In detail, Pearson (1998) points out the qualitative criteria characterizing texts, (i) texts need to be recent and current, this means that it is advisable to design a synchronic corpus which is determined by texts reflecting the ongoing domain status, in this way it can be possible to provide lists of domain-specific terms used in a given moment alongside updated conceptual information; (ii) texts have to be published by several experts, such as, institutions, domain experts, organizations, etc.; (iii) texts should ideally be in their original form, so not translations because this is the only way that guarantees the terminology being accurate and original in any specialized areas; (iv) texts inside corpora have to be integral avoiding the fragmented proportions of their content in order not to hinder the contextualization of the information. Moreover, if the terminologist's work is intended to represent specific geographic sectors, it should rather fix some placing borders to carry out text mining passages. Given these specific overall criteria, the trustworthiness of the included texts in a corpus should be achieved. Some years earlier than Pearson,

Biber (1993) [44] outlined the design of the corpus as a preliminary phase from which to begin the more accurate extraction process of the information included in a specialized corpus. In his work "Representativeness in corpus design," the author highlights the features that characterize the corpus configuration procedures, i.e., the selection of texts and the size of these sampling frames aiming at assuming that a corpus, in order to be representative of a domain, should reflect its lexicon in a comprehensive way. Among the theoretical approaches proposed by Biber's corpus-based design phases, there is the one on how he overcomes the quite common idea of reaching a representativeness by only considering the numerosity of texts included in it. He rather suggests to keep in view the target population, to which the corpus processing has to be used to reach some cultural and scientific needs, as a main feature in measuring the coverage. Which is the extent of an ideal or potential population comes to be considered as a significant factor towards the achievement of information completeness. When explaining several perspectives to compute corpora representativeness, the author mentions the variability feature in texts *genres*, i.e., a distributional variation in a language proper to the corpora under development, as to provide basic principles upon which to support the analysis of terminological adjustments according to concrete application cases. This means that a corpus is supposed to represent the different text typologies and the terminology contained within it, which has to be kept updated, specific to a given population, otherwise the distributional value of its lexical content can hardly be reached. What characterizes the linguistic coverage is, still taking into account Biber's position, the number of words in the texts as well as the several typologies of texts in a certain sample. Given for granted the constraints found in determining which absolute measures can be considered the best threshold scores according to which a corpus should be treated as fully covering a domain, the phase concerning the selection of texts plays a crucial role in revealing the target population and the usages expectations for which the corpus is meant to be applied on. Stratification of texts is, among the others, one of the solutions Biber points out as supporting the corpus design development. According to this principle, the author stated

> "[...] stratified samples are almost always more representative than non-stratified samples (and they are nevertheless representative). This is because identified strata can be fully represented (100 percent sampling) in the proportions desired, rather than depending on random selection techniques. In statistical terms, the between-group variance is typically larger than within-group variance and thus a sample that forces representation across identifiable groups will be more representative overall." (1993:247)

Texts can be selected by filtering out several features, as for example by considering the genres, the topics and the sources. To support the decision-making process of texts inclusion in the source corpora and then to disseminate the terminological resources, the development of specific and technical collections proves to be crucial as a starting point from which to begin the text processing. As affirmed by Barriére (2006) [5], a corpus compilation is the first passage essential towards the construction of the terminological knowledge bases (TKBs), which will be effectively the point from which to create the semantic resources.

1.2   Text mining

To discover new information about a field of knowledge whose semantic components or communicative scope are to be analyzed for deeper digital evaluations, text mining is an efficient technique that enables users to extract information from huge amount of datasets [117]. Executing text mining procedures involves a series of operations that facilitate end users in reaching unstructured range of data and applying to them some techniques of categorization or retrieval of certain textual parameters. In the literature many authors have provided several examples of text mining techniques, for instance, according to Maheswari (2017) and Dang (2015) [51], the following are the main ones: clustering, text categorization, text summarization, sentiment analysis, Talib (2016) [153] also considers information extraction and NLP (specifically related to Named Entity Recognition –NER–), while Salloum (2018) [160] adds also *Association rule extractors*, particularly, K-means algorithms, word clouds. In detail, when authors refer to text mining techniques, specifically they describe them as grouped in six main categories:

■ Information extraction: authors agree in considering the extraction of information for specific communicative purposes a methodology that extracts specific data from unstructured frameworks to be further used in search engines operations.

■ Text Clustering, or Association rule extractors: among these techniques, which classify several documents into classes through unsupervised methods, one of the most used ones is the *K-means* algorithm [139], which subdivides the source data bank into clusters, specifically k cluster, each one of them is represented by a centroid. The execution of this algorithm is cyclic, meaning that it ends up until the k cluster centroid reaches a constant status approaching every time to the nearest centroid.

■ Text categorization: following the words of Dang (2015:2465): "The categorization is supervised process and uses predefined set documents according to their contents. Responsiveness and flexibility of the post-co-ordinate system effectively prohibit the establishment of meaningful relationships because a category is created by individual not the system. While as the clustering is used to find intrinsic structures in information and arrange them into related subgroups for further study and analysis.

It is an unsupervised process through which objects are classified into groups called clusters." Among the main algorithms used to represent the categorization in text mining there is the *Support Vector Machine (SVM)* [112], the *K-NN (K-Nearest Neighbor Classification* [175], the *Decision Trees* [35].

■ Text summarization: is the process of reducing a text in its salient components, relevant to further semantic studies; in this procedure several algorithms could be exploited to extract and create condensed models of the texts, e.g., *KeyPhrases extractor* [51]; *SentenceRank* [21] which creates a graph made up of nodes, these latter "are considered as sentences and edges are semantic relatedness between that sentences which is calculated by with the help of WordNet. Rank these nodes using a ranking algorithm and select top ranked sentence as summary" (Maheswari, 2017:1663); or *LexRank*, "based on the concept of eigenvector centrality in a graph representation of sentences. In this model, a connectivity matrix based on intra-sentence cosine similarity is used as the adjacency matrix of the graph representation of sentences" (Erkan, 2004: 457)[14].

■ Information visualization, in the form of word clouds [160], are visual illustrations of the words having some levels of proximity with each other in the texts given in input through statistical measurements.

1.2.1 Data pre-processing

Prior to executing text mining techniques, texts should be pre-processed. Text mining consists in a series of extraction procedures that can reveal the linguistically and statistically significant terms representing domain corpus concepts [134]. To reach a high level of proximity within terms to key concepts of the area of study, as Pazienza (2005) underlines, there are three main approaches to undertake, i.e., linguistic and statistic, or the hybrid one that merges the previous two. The linguistic approach to term recognition relies on the preliminary steps to clean the texts to be processed in order to achieve in output a list of candidate terms of a domain. These phases are usually related to the *parsing* operations with respect to PosTagging (POS) analysis for the language to be studied – in this research study the languages taken into consideration are first Italian and then English – in order to group the lexical units onto which apply regular expressions rules; successively, another step deals with the removal of parts of text that are not meaningful, such as, adverbs or articles, the so-called stopwords, and then it is possible to detect preferred terms that can be considered as part of candidates to be integrated in the semantic resources. One feature to be considered is the variation of terms, the several forms with which a term can be shaped and how these can be conceived or not as triggers to develop a hierarchical or synonymous semantic organization. In case the variation in terms leads to single standing terms, they can pass as candidate term to be evaluated by experts; on the contrary, terms can provide semantic connections based on their varied forms to be studied to create a terminological network system. Once these first operations on texts have been accomplished, the resulted candidate terms should be examined by a group of domain experts, which is meant to give an objective evaluation on the appropriateness of the lexical units and their semantic connections. Pazienza (2005), in this respect, draws the attention on the effective support of statistical measures attributed to each of the single or complex terms of domain corpora whose scores should be held in consideration to understand their unithood with domain concepts meant to be terminologically represented, and thus be perceived as representative terms carrying unique and authoritative meaning. Indeed, he underlined that:

> "Statistical measures applied to terminology are of great help in ranking extracted candidate terms according to a criterion able to distinguish among true and false terms and able to give higher emphasis to 'better' terms. What is expected an ideal statistical measure could do is to assign higher scores to those candidates supposed to strongly possess a peculiar property characterizing terms." (2005:5)

In the linguistic analysis of extracted terms of a particular domain, one of the most reliable statistical measures able to show the level of unithood [97] terms hold, is the frequency measurement, which indicates, by discarding a series of not meaningful lexical units and filtering out a list of single or multi-word terms that the more a term is in high positions in the ranking, the more it is representative of a specialized concept of a domain. This principle has been followed for the purposes of this book, as the further Table 4.2 on the levels of specificity with respect to generalized corpora will demonstrate. Terms frequency measures take ground from statistical evaluation of term candidates inclusion from a term extraction list, mono and multi word term banks, parsing of verbal expression and categories filtering for NP-constructions [69]. The relevancy of terms extracted depends, following Thurmair's work (2003), on the coherence with the domain they should represent, or when it comes to regards information retrieval,

> "In the case of retrieval, relevancy is defined by the searchable concepts of a corpus." (2003:4)

One of the most exploited methods in computational linguistics to evaluate the relevancy of the term extraction is the Recall and Precision [38] formula, even if it does not fully explain on which levels the relevant terms should be considered to be selected. Specifically, Recall represents the number of correct terms detected in the term extraction lists divided by all the correct terms contained in the gold standards, while Precision refers to all the candidates detected (positive and negative) divided by all the correct terms in the gold standards. Cabré (2001:3) gives a explanation of these measurements:

"The two most frequently used measures in the assessment of these systems are found in IR: recall and precision. Recall is defined as the relationship between the sum of retrieved terms and the sum of existing terms in the document that is being explored. In contrast precision accounts for the relationship between those extracted terms that are really terms and the aggregate of candidate terms that are found. These measures can be interpreted as the capacity of the detection system to extract all terms from a document (recall) and the capacity to discriminate between those units detected by the system which are terms and those which are not (precision)."

Still in Thurmair (2003:5), we can find a list of characteristics relevancy should present depending on the contexts:

■ "In terminology, relevancy of a term is determined by its position in an LSP domain;

■ in translation, relevancy of a term is determined by its status vis-à-vis an existing term bank or dictionary, the entries of with have an unclear status and coverage themselves;

■ in retrieval, relevancy of a term is determined by the possibility to search for it, which holds for basically all terms of the corpus."

The appropriateness of terms as results of a semantic automatized extraction has also been described in Cabré's review on terminological extractor tools [128], specifically referring to the constraints these latter face in their practical application. In particular, what is firstly highlighted by the author is the weak level from these tools in the ability of detecting the terminological nature of lexical units, being them belonging to general areas of study or specialized ones. As stated in Nakagawa (2002) [76], candidate terms are selected after having applied a filtering on source corpus texts of their *stop word lists* made of expressions like "*of*," "*the*". What is extracted is indeed a list of single or complex terms, which can be collocations resulting from the simple word units of the extracted terms, and they should present a high level of *unithood*, a concept introduced by Kageura (1996) meaning the degree of strength of syntagmatic collocations.

1.3   Terminological knowledge bases (TBK)

The intent of this study is to provide an Italian resource, firstly conceived as a thesaurus and then converted into an ontology, to configure the terminology of Cybersecurity in a network of semantic relations that can better orientate to a terminological understanding of specialized concepts represented by terms belonging to this field of study. Terms included in the thesauri have to keep an unambiguous value, as affirmed in the standard NISO TR-06-2017, Issues in Vocabulary Management: "Controlled vocabulary: A list of terms that have been enumerated explicitly. This list is controlled by and is available from a controlled vocabulary registration authority. All terms in a controlled vocabulary must have an unambiguous, non-redundant definition." Constructing an efficient terminological system usually implies the acquisition of domain-oriented information from texts, specifically those that can provide semantic knowledge density and granularity about the lexicon that is meant to be represented [42]. These structures are in literature known as Terminological Knowledge Bases (TKBs) [13]. In detail, they support the modalities of systematizing the specialized knowledge by merging the skills proper to linguistics and knowledge engineering. The purpose of creating solid terminological sources, through a selection of candidate representative terms, is linked to the attempt of providing a reliable reflection of the specialized knowledge included in source corpora [87]. Condamines (2018) stresses the importance for TKBs of being corpus specificity aligned, in the way they should adhere to the unambiguous nature of specialized domains as well as to the application contexts in which they are meant to be used [92]. The ways in which the candidate terms are extracted from a specific domain-oriented corpus [56] usually follow text pre-processing procedures and extraction of single and multi-word units [32] from texts filtered out by frequency measures, then they can undergo a phase of variation recognition [125] and other statistical calculations to determine the specificity, accuracy and similarity in the texts from which they come from [128]. The reason why the domain-oriented terms are called "candidates" [13] is linked to the fact that in the terminologists' activity the need of experts' validation is frequently required, this is because just the subjective selection by terminologists might not be exhaustive and fully consistent with the domain expertise (ISO/TC 46/SC 9 2013, ISO 25964-2:2013).

Thesauri construction is commonly connoted by a manual semantic work that assumes a terminologists' activity in selecting terms from a list of candidates ones, extracted, in turn, from a reference corpus [11] and, consequently, arranging them in a structure that follows the guidelines given by ISO standards for structuring thesauri [2], [83], which aim at normalizing the information meant to be shared by a community of users. To make terminologists' process of defining thesauri's structure [119] less hard, their construction phases are supported by using computer engineering techniques and followed by an evaluation phase that sees experts of the domain involved in the decision-making process about the insertion of the terms in the semantic resource. Nonetheless, the process of appropriateness' check by experts is not entirely suitable to demonstrate that the TKBs comply with the specialized corpus knowledge flow. Hence, together with specific groups of experts' supervision, other tools support the accuracy validation, that is existing terminologies mapping systems dealing with the specific areas of interest officially accepted and employed by the sector-oriented working groups, i.e., the gold standards [42]. This task is meant to give results on the way terms, that have been selected to be part of a semantic resource – designed to represent a specialized language –, can be aligned with the ones included in reference authoritative texts. These target texts can be in the same language as the one of the source corpus, and could present less difficulties in the matching system, or multilingual [22], in these cases using translations from existing semantic resources could represent a solution. In this paper, the gold standards taken into account are in Italian language or have been translated in Italian – NIST 7298 [102] and ISO 27000:2016 [82] – this reflects the native purpose of the project that was intended to provide a guidance for the understanding of the Cybersecurity domain in Italian language. The process of creating reliable TBKs is particularly associated with the synergy of terminology and knowledge engineering fields [13]. The terminology tasks in the phases meant to create semantic resources to be applied to specific user cases in technical domains is strictly supported by NLP techniques and other knowledge engineering operations, such as, clustering or indexing [124]. Still in Condamines (2018), there is a detailed list of advantages terminology and knowledge engineering systems respectively provide. For instance, the linguistic role given by terminology can guarantee an accurate way of filtering out concepts and their relation networks, but, on the other hand, it is just by having the support of computerized techniques that terminologists can properly analyze big corpora and detect linguistic elements to include in the semantic tools. The progresses in the study of the candidate terms of a specialized domain in order to insert them in a system that can orientate its comprehension by the users are closely related to the automatic extraction from texts composing the source corpora.

1.3.1 *Automatic term extraction (ATE)*

As specified by Cabré (2001), term extraction can deal with two different purposes: on the one hand, it helps in gathering all the terminological units inside source corpora – and that is crucial to the development of a series of resources, such as controlled vocabularies, ontologies –; on the other hand, it supports the Information Retrieval (IR) phases in collecting just the lexical units that can be used as indexing seeds for documents selection, and, thus, helping out the interaction between users and digital contents for specific needs. Zadeh (2014) [145] pointed out the two steps of the terminological identification, (i) terms extraction and (ii) sorting their frequency scores. Terms identification results to be important both to extract specialized terminology proper to domains of study and to perform syntactic analysis towards hierarchical frameworks construction [24]. Alongside the proposed mixed approach as Aubin (2006), the first phases undertaken in this study are based on the terminology extraction from the Cybersecurity specialized corpus of documents following statistical measures, reference to external gold standards and experts supported task of term filtering from linguistic and communicative perspectives. More specifically, it is Aubin who presents

Yatea as software that extracts Noun Phrases and their syntactic analysis in a "head-modifier format" [24]. The way by which it executes terminology operations are described by the author (2006:4):

> "[...]As an input, the term extractor requires a corpus which has been segmented into words and sentences, lemmatized and tagged with part-of-speech (POS) information. The implementation of this term extractor allows to process large corpora. It is not dependent on a specific language in the sense that all linguistic features can be modified or created for a new language, sub-language or tagset. The main strategy of analysis of the term candidates is based on the exploitation of simple parsing patterns and endogenous disambiguation. Exogenous disambiguation is also made possible for the identification and the analysis of term candidates by the use of external resources, i.e. lists of testified terms. YATEA allows exogenous disambiguation, i.e. the exploitation of existing (testified) terminologies to assist the chunking, parsing and extraction steps. During chunking, sequences of words corresponding to testified terms are identified. They cannot be further split or deleted. Their POS tags and lemmas can be corrected according to those associated to the testified term. If an MNP corresponds to a testified term for which a parse exists (provided by the user or computed using parsing patterns), it is recorded as a term candidate with the highest score of reliability."

Thurmair (2003), pointed out that the extraction of terms in a domain of study consists in two passages: one involving the proper extraction of terms from specialized corpora, and the other dealing with the identification, and selection, of the resulted terms compared with the existing external resources to keep an upper certified level of shared knowledge. Automatic term extraction is based on several analysis of corpora, and, as underlined by Cabré (2001), three main approaches to detect terminology from specialized domains can be pursued, (i) linguistic, (ii) statistical or (iii) hybrid, merging the linguistic analysis together with the statistical measures [134]. Whichever perspective terminologists undertake, the results of semantic extraction are related to electronic data sets and the term lists given in output by the automatized tools are characterized by several columns displaying their frequency scores in the source corpora and the semantic context to infer knowledge. Selected terms as preferred entry lexical units representing specialized domain concepts are usually selected according to the *Mutual information* level they have, which is a term occurrence over a certain level. The preliminary steps needed before the development of a semantic resource meant to be applied by experts from a particular domain are the compilation of a specialized corpus and the terminological extraction from it. The documents gathered in the corpus undergo a process of semantic extraction that is going to pre-process the texts and give as output a list of the most representative candidate terms meant to be part of the terminological tool. The terms resulting from the corpus processing shall be compared in the perspective of presenting the semantic coverage with the representative ones included in the main domain-oriented gold standards [22].

The terms of the list resulting from the reference corpora processing passages are usually given as output with the implementation of semi-automatic term extraction tools, which use a large range of NLP techniques in order to provide controlled terminological lists to be analyzed by experts of a specialized domain. Nazarenko *et al.* (2009) [19] and Loginova (2012) reported detailed lists of several tools for extracting terminologies from texts. With regards to the Cybersecurity domain and the research activity dealt within this book, various existing sources, both in English and in Italian, have been analyzed in order to retrieve an accurate terminological baseline from which to start to build a more fine-grained semantic resource to guide the knowledge representation process [113]. As stated in Kosa (2017) [172], the saturation of a term list representing a domain can guarantee a knowledge coverage of the area of study, and by doing so it is important from a terminologist's point of view to understand to which extent the software tools for automatically extracting terms from source corpora can help in reaching this goal. Even though, as Vazquez (2018) stressed out, domain-oriented terminologies detected by exploiting a set of NLP techniques through term extractor tools have to be evaluated by specialists of these sectors and manually selected in accordance with the shared technical knowledge to calculate their coverage. It is for this reason that the semantic tools that should represent a way to monitor a specific terminology and guide the understanding of the key component concepts of an area of study, such as thesauri or ontologies, have to be compared in their terms lists, as well as their semantic relations connections, against external reference resources officially recognized as reliable and authoritative comparison models.

1.3.2 *Gold standards reference*

Evaluation is a crucial phase to infer if a terminology bank can or cannot be suitable to represent a specialized domain [134]. Many statistical approaches have been exploited in the literature to compute the percentage extent terms show in their relevance proximity to domain corpus documents, i.e., (i) Frequency measures (Term Frequency, Inverse Document Frequency, Term Frequency–Inverse Document Frequency); (ii) Mutual Information (computes the information level obtained on a variable by observing the other, it reduces the uncertainty [100]); (iii) Loglikelihood score, a statistical test to verify the interdependence hypothesis between two words in given corpora; (iv) CNC-Value that aims at providing a better extraction of nested multi-word terms and integrating context information [96]. Even though these calculations are necessary to have a clear overlook of terminological frameworks for a domain given certain best candidate terms selected according to their ranking position in the list resulting from term extractions, a phase that still seems significant is that referred to manual experts checking process. This evaluation, prior to the inclusion of meaningful terms for the domain to be represented, is carried out through an hybrid approach that sees the linguistic procedures, allowing a post-processed document information, merged with statistical computation metrics, which, on the other hand, guarantee a fixed measure to prove how the selected terms are relevant in determining domain-specific concepts. To achieve this latter goal, considering external authoritative resources, *gold standards*, can be helpful in filtering out in a better way the terms that are officially more precise with respect to the existing ones in the literature [5]. In this regard, this book is focused on taking into account a comparative analysis with some of the main gold references for Cybersecurity domain of study in order to ensure the highest level of reliability as possible for the Cybersecurity thesaurus and ontology knowledge representation. The measures employed to level up the selection of true and false terms are Recall, Precision and F1-measure, as well as Tf*Idf and Domain Relevance.

1.3.3   Domain-dependency

As stated by L'Homme (2014) [113], terms are strictly dependent from the contexts in which they are supposed to be employed. Which field will be covered represents a key feature to take into account once terms will be chosen to represent conceptual information about a domain. To better clarify this relevant bond existing in deciding the specificity of terms given a certain domain, in the Cybersecurity field of knowledge, for example, *virus* implies a series of assumptions this term brings with itself in representing the conceptual model that proves to be different from the same terms used in a medical context. Understanding the flexibility provided by the application of preferred terms in precise technical areas of study is highly significant in the meaning these latter will provide to common users as well as experts. As a result, it seems quite crucial to find a way to avoid ambiguity and polisemy among terms because they could lead to different senses relating to a conceptual model. Exploring works in computational terminology such as the one carried out by Kageura and Umino (1996) [97], a common trait terms present when they come to be linked to very specific conceptual meanings according to the domain they are representing, is their *termhood*, that is the stable lexical connection they exhibit with respect to domain-specific concepts. In order come to a form of agreement on the univocal employment of specific terms over a specialized domain, the standardization results to be indispensable.

1.3.4   Terms population

In addition to thesauri extent, also for terms, the criterion related to representativeness can be a significant measure to establish if the resources which gather the information of a specialized domain can be taken as a paradigm. While for thesauri construction attention should be given firstly to the composition of documents making up the source corpus to be processed, the terms analysis usually relies on a post-processing phase where they should already be lemmatized and tokenized as to facilitate the selection step for the preferred ones. The feature that distinguishes the most representative terms representing a domain concepts is the frequency value associated to the text extractions on corpora documents, this can be measured through IDF or TF statistical calculations [158]. This parameter is helpful in outlining the first most common and specific terms on the top of the source corpus files. The specificity degree can be computed by contrasting the specialized corpus with a generic one, this can highlight the weight of terms in different contexts and help the selection of the most technical ones for the domain [165]. The semantic resources containing the representative terms that reflect the conceptual framework of specific domains of study are meant to be updated in the perspective of following the variability in the vocabulary. In Chapter 6 a RDF graph (6.1) will show a perspective of an alerting system according to which through a triple of blank node it can be possible to receive a notification of specific document integration that can be the trigger to reconsider a semantic structure of a means like a thesaurus and ontology, and by consequence a starting point from which to add new terms or substitute existing ones with others that have become more popular in the specialized groups of domain experts.

Chapter 2

Case study: Cybersecurity domain

2.1 OCS project

This research project is a joint collaboration with the Cybersecurity Observatory (OCS) at the IIT-CNR[1], which has as its main purpose that of constructing a platform of information to intensify the perception of importance the security in informative systems holds within specific areas, such as enterprises, professionals and public audience. Many services have been developed inside this web-based platform which involves:

- several identification reviews on vulnerabilities in informative systems;
- threats detection;
- self-assessment forms;
- tweets analysis;
- exploits analysis;
- Domain generation algorithms (DGA) tracking;
- ransomware recording;
- 3D attacks map;
- antimalware report;
- thesaurus;
- ontology;
- e-mail spam detection;
- General Data Protection Regulation (GDPR) tools.

---

[1] https://www.cybersecurityosservatorio.it/it

DOI: 10.1201/9781003281450-2

The OCS is supported by CNR, the European Commission, the Registry.it, the Tuscany Region and other entities. Its updated news are accessible to the public with the objective of increasing the resilience against Cybersecurity malicious events as well as a sense of awareness for all the actions that should be undertaken to rapidly react to the cyber attacks or pursue a backhand strategy. Moreover, providing many analyses about the latest trends in the attacks and in the threats occurring in many informative working environments, all the parties interested will have a reliable and detailed framework to take into account in carrying out specific actions in their sectors. For this purpose, being a platform that aims at representing an official and trustworthy means for Cybersecurity information dissemination, the sections are accompanied also by a connection page with different representative legal documents about Cybersecurity in Italian and in English, including the best practices, guidelines, national frameworks and standards.

This study addresses the creation of a thesaurus in Italian language that can be employed as a semantic model towards the management of the Cybersecurity specific terminology, and then the migration of the structured information into an ontology system following the rules proper to RDF and OWL languages to be able to ensure a higher level of interoperability with other informative systems. The OCS online platform represents a collaborative activity together with Cybersecurity experts of IIT-CNR in Italy, which involves several sections having as primary objective that of providing as much updated information as possible about the domain and a baseline supporting the defence activities procedures. The thesaurus and the ontology are integrated amongst the platform services subdivisions. The main goal of this semantic branch has been that of enabling users to get closer to difficult concepts related to the domain of Cybersecurity exploiting the properties of both thesauri and ontologies in organizing and representing a specialized field of knowledge. Besides the help for common users in more clearly identifying some key concepts of the domain, these semantic tools have been taken into account as part of the OCS services for their correspondence with authoritative documents from which the Cybersecurity informative tissue comes from, i.e., NIST 7298 or ISO 27000:2016. Hence, the thesaurus, and then the ontology, have represented a computational linguistic way to enable users to explore, learn and investigate the main representative terms interrelated with others of the domain by hierarchy, equivalence and associative connections, with the result of displaying an overall semantic integrated and structured network to orientate the domain perception [95]. The web page of the OCS project has been developed with the goal of providing reliable sets of domain-oriented information for the Italian experts as well as public users. The statistics regarding the visits by the users to this web Cybersecurity service are based on accesses from several countries, this feature reveals the international significance the platform holds. Data show an higher number of Italian users who have accessed to be updated about the news of the latest Cybersecurity defence strategies, as can be observed by looking at Figure 2.1. The interest shown by the Italian audience

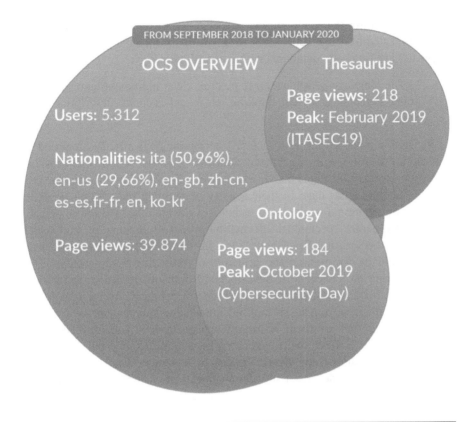

Figure 2.1: Statistics OCS website

proves that this resource has specifically been realized for an Italian context and public, as well as for the concrete end-users applicability. The diagram depicts the generic scores of visits on the main web pages and, in particular, numbers referred to the thesaurus and ontology accesses. Even if no outstanding statistic scores related to semantic tools visits can be observed, it is possible to underline that the peak level of users who looked up terminological services has been reached during two events where the services have been launched to a heterogeneous audience made up of enterprise professionals, stakeholders, third-parties service providers, experts of the domain, academics and common users.

2.2 Specialized languages reflecting specialized domains

As languages evolve over decades, also the technical information proper to the languages characterizing several fields of study should reflect this semantic variation including the inner improved specificity to represent their concepts [133]. As stated by Nagy (2015:265) [81], specialized languages are those that enable to have linguistic databases with univocal meanings:

> "The language of science(s) is precise, clear and unambiguous. Impersonal statements, logical thinking, clear and accurate descriptions prevail[...]"

In Pavel's monography (2002) [156], terminology is strictly identified as a discipline dealing with specialized domains of study which organizes unambiguous terms and concepts. In particular, the author highlights the crucial role teminologists have in their activity of merging their linguistic skills with the computational engineering techniques towards the creation of semantic tools to organize specialized information proper to domains of study, such as that of Cybersecurity. Pavel underlines the fact that the terminological research should be focused on the principle of providing a representation of terms reflecting the concepts of the structured domains of specialized knowledge meant to be a vehicular dissemination of technical information to the public. Domains referring to specialized knowledge spheres can be found in all sectors, and each of them is usually represented by a controlled list of terms that can be a thesaurus, a taxonomy or a more detailed ontology to allow its understanding from diverse ranges of audience. The interconnection among terms is provided by a set of semantic relationships that enable users to have an overview of the information in a structured way. For instance, dealing with the *medical* area of study, MeSH (Medical Subject Heading)[2] thesaural system by the National Library of Medicine to index and catalogue biomedical information might be mentioned – it has also been converted into an ontology [120]; or for what regards the agriculture area, it is possible to publicly consult the thesaurus AgroVoc by the Food and Agriculture Organization (FAO) of the United Nations[3]. Both MeSH and AgroVoc thesauri are multilingual and they help out indexers and cataloguers in structuring the technical information that is organized in an hierarchical way and completed by synonyms and associations with the other terms reflecting the concepts of the respective domains. Figure 2.2[4] shows the tree view MeSH thesaurus provides to help indexers in querying specific terms or subjects.

It appears to be essential for semantic tools that are developed to represent key concepts of specialized domains to draw on reliable documents from which to retrieve useful information. The latter in order to be transformed in structured data, which might be managed by a semantic tool that follows standardized guidelines in the way it organizes domain-specific terminologies, should then be treated through several steps of terminological processing [58] with the

---

[2]https://www.ncbi.nlm.nih.gov/mesh/
[3]http://aims.fao.org/vest-registry/vocabularies/agrovoc
[4]https://meshb.nlm.nih.gov/record/ui?ui=D014780

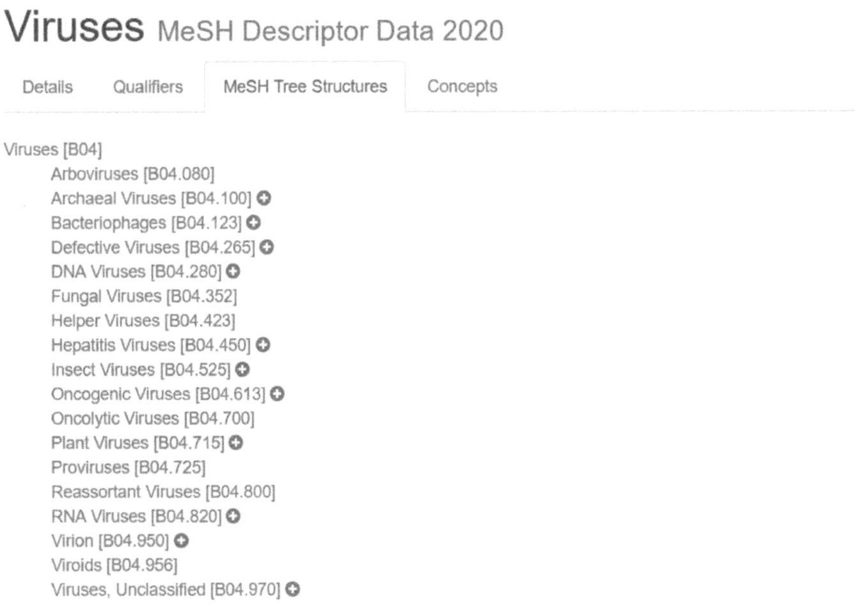

Viruses MeSH Descriptor Data 2020

Details   Qualifiers   MeSH Tree Structures   Concepts

Viruses [B04]
   Arboviruses [B04.080]
   Archaeal Viruses [B04.100] ◐
   Bacteriophages [B04.123] ◐
   Defective Viruses [B04.265] ◐
   DNA Viruses [B04.280] ◐
   Fungal Viruses [B04.352]
   Helper Viruses [B04.423]
   Hepatitis Viruses [B04.450] ◐
   Insect Viruses [B04.525] ◐
   Oncogenic Viruses [B04.613] ◐
   Oncolytic Viruses [B04.700]
   Plant Viruses [B04.715] ◐
   Proviruses [B04.725]
   Reassortant Viruses [B04.800]
   RNA Viruses [B04.820] ◐
   Virion [B04.950] ◐
   Viroids [B04.956]
   Viruses, Unclassified [B04.970] ◐

Figure 2.2: MeSH tree view for virus

use of knowledge engineering skills and computer-assisted semantic automatic tools. Indeed, it is by the means of distributional semantics (patterns configuration, word embedding structures) that the relationship systematization for the Italian thesaurus for Cybersecurity has been possible to be achieved. Therefore, the composition of the source corpus represents the preliminary phase which enables the realization of a means of knowledge organization and representation to classify and systematize its inner technical information.

2.3   Cybersecurity field of knowledge

The Cybersecurity domain is strictly linked to many sub-fields that intact a range of social environments supposedly subjected to computer attacks. The digital transformation has implied over the years a new way of conceiving the protection barriers against *in-loco* and external attacks. Cybersecurity, and what is named *cyber hygiene* [147], provides strategies that can repeal the effects of malicious threats to the informative systems. As described in the White Book on Cybersecurity (2018:3):

"In an increasingly digitalised world, cyber attacks are alarming popula-
tions, causing huge damage to the economy, and endangering the lives of
citizens when they affect distribution networks for essential services such
as healthcase, energy, transport, i.e., critical infrastructures of modern society. [...] a successful cyber attack could also represent the point of no
return for a company's credibility, the development of its business and its
ability to competitevily sell products."

Indeed, as further analyzed in the aforementioned Italian handbook for Cybersecurity, the pervasiveness of cyber risks occurs in many sectors, e.g., with the
episodes of identity thefts while giving personal information during the *Smart
Tourism* events on mobile devices apps; or a power failure in electronics; pay-
ments through banking systems that can be automated or replicated; *phishing*
attacks in personal or working e-mail addresses; self-driving cars that can be re-
motely controlled or cracked; IoT technologies in industrial environments relying on cloud; blockchain technologies and artificial intelligence can be damaged
by losing critical information; alarming terms used in press and communication frameworks that can provoke a misunderstanding of many concepts related
to Cybersecurity strategies. Many of the attacks exploit the vulnerabilities of
the systems and the *human factor* that can accidentally lead to other malicious
threats. For this reason, arranging a set of sound techniques to ensure the invi-
olability of personal data can become essential. Cybersecurity allows to apply
several methods to overcome or prevent malicious attacks, firstly linked to personal data protection through biometrics methodologies, e.g., face recognition,
finger prints, irises, voice and so on [147], and others methods referred to network protection techniques. One of the main portals upon which it is possible
to find the latest lists about the attacks techniques as well as the mitigation procedures is the one published by MITRE, both with the ATTCK webpage, which
subdivides every description dealing with enterprises or mobile[5], and through
the platform that tracks the vulnerabilities of the systems, the so-called Common
Vulnerabilities and Exposures (CVE)[6], by specific unique codes able to provide
unambiguous information about the most frequent vulnerabilities the enterprises
can have and which can lead to severe cyber disruptive attacks for the organization. For what regards the statistics about the latest trends of cyber threats, or
major data breaches which have caused huge financial losses, NIST web portal
provides an enumeration of the most visited and analytic data[7]. Being permeated
by a synergy application over various technical sectors, the Cybersecurity field
of knowledge proves to be symbolized by a high technical terminology structure
coming from several areas. Its cross-curricular framework integrates computer

---

[5]https://attack.mitre.org/techniques/mobile/
[6]https://cve.mitre.org/find/index.html
[7]https://www.nist.gov/blogs/manufacturing-innovation-blog/20-cybersecurity-statistics-manufacturers-cant

technologies as well as electronics, or legislative contexts that regulate the different country section management of cyber risks, and standardized shared terminology to represent key concepts to experts and common users. It is specifically on this last aspect that this research study is focused on, with the objective of guaranteeing an homogeneous tool, commonly approved by domain experts, to cover the specificity and variation of the Cybersecurity domain lexicon [93].

2.4 Corpus design for Cybersecurity domain in Italian language

The Italian thesaurus for Cybersecurity, built in collaboration with IIT-CNR for the OCS project, has been developed starting from a range of trustworthy documents published by authoritative institutions and organisms. To realize such a reliable set of documents to be consulted for the information processing, the hierarchy of sources in law has been followed [71]. This classification system for the legislative types of texts to refer to as reference documents organizes the order from which to start the inspection of these latter that should create the baseline for the source corpus in macro-subject groups, as shown in Figure 2.3.

Hence, the Italian thesaurus has been based on this sequence, starting with legislative sets of documents, retrieved by European, and in the multilingual conversion, international, official sources, from public web pages, such as EUR-lex, Altalex. These web-based consultation platforms enable users open access to documents of interest in specific areas of economics, politics, health systems, academics from a legislative perspective. Several filters have been applied in the information retrieval system, for example, with regards to the classification of legislative sources (Penal, Civil, Administrative Code), time, keywords as query items to search for information within the broad databases, as section 4.1 will describe in detail. In this way the set of laws that constitutes the source corpus has been populated by the main Cybersecurity legislation from an European perspective, in particular the Italian framework as a start. Among the most important and considered as a benchmarks in the Cybersecurity law systems, the source legislative corpus included the (i) EU NIS Directive (Network and Information Security) in force starting from August 2016, it covers some of the key issues referring to the interoperability concerning security strategic plans among the EU Member States or to the reports for the risk management, specifically it aims at encouraging a communication strategy in order to reveal crisis caused by risks or security failures within organizations and political informative environments; (ii) the Data Protection law, known as GDPR, which became applicable from May 2018; (iii) the decree of the Prime Minister in force from 17th February 2017, publicly known as Gentiloni's Decree in Italy, which sets out a crisis management plan focusing on the empowerment of some security

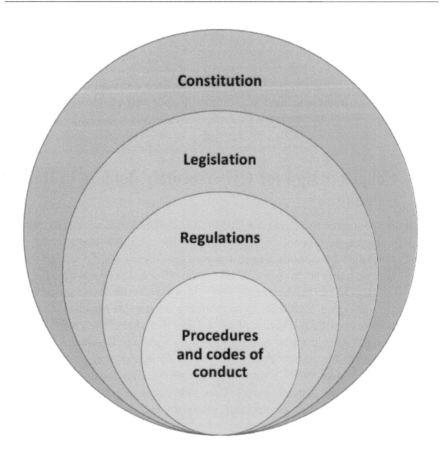

Figure 2.3: Hierarchy of laws

bodies [147] *Security Intelligence Department (DIS)*[8], *National Securituy Board (NSC)*[9]; other important laws on the Cybersecurity domain will be presented in Chapter 4 in the description of the source corpus composition.

In this regard, it is worth mentioning here some of the official organisms that are in charge of security strategies implementation for political institutions and organizations:

■ the European Union Agency for Cybersecurity (ENISA)[10];

■ the Italian risk prevention and incident response group **– Computer Emergency Responce Team** (CERT), which from May 2020 has been merged into the CSIRT in Italy for the Cyber Defence;

---

[8]https://www.sicurezzanazionale.gov.it/sisr.nsf/chi-siamo/organizzazione/dis.html
[9]https://www.sicurezzaegiustizia.com/istituito-il-nucleo-per-la-\\sicurezza-cibernetica/
[10]https://www.enisa.europa.eu/

■ the international bodies **Computer Security Incident Response Team** (CISRTs)[11], which are aimed to report incidents through an alert system platform, prevent attacks before the actual occurrence and provide a risk analysis exposure report;

■ the Italian **Committee for the Security of the Republic** (CISR)[12];

■ the **National Cybercrime Centre for Critical Infrastrucutre Protection** (CNAIPIC)[13];

To better clarify the composition of the main representative groups in the security field of knowledge, the White Handbook for Cybersecurity provides an explanatory diagram to help in understanding the architecture behind the Cyber Defence bodies as shown in Figure 2.4.

Apart from the legislative documents which have characterized the authoritative section of the source corpus for the Cybersecurity thesaurus realization, being objective of this book the monitoring supervision for terminology variation exploiting the knowledge engineering skills, the domain spectrum has been enhanced with other types of texts. In the following subsection, a list of existing and accessible resources that have been considered as part of the corpus and reference texts for the mapping systems will be further described. For instance, these are guidelines, domain-oriented magazines, standards and frameworks.

2.5   Existing resources for Cybersecurity domain

In this section a list of bilingual existing online sources, with open and sometimes restricted access for public users, of information about the Cybersecurity domain is provided [129]. The official sources of specialized knowledge are considered as a set of documents published by authoritative figures or institutions that can guarantee a reliable level of specific domain-oriented terminology to be used to build up the thesaurus and the ontology for the Cybersecurity area. The following existing sources are meant to be conceived as gold standards to be used in the mapping alignment with the thesaurus and the ontology to provide an accurate coverage correspondence of the domain [63]. In particular, the Italian sources will represent the *parallel* corpus [52] to which to refer to in order to verify the degree of specificity the thesaurus is showing with respect to the officially approved and shared terminology on Cybersecurity; the English documents, in

---

[11] https://www.enisa.europa.eu/topics/csirts-in-europe/csirt-inventory/certs-by-country-interactive-map
[12] https://www.sicurezzanazionale.gov.it/sisr.nsf/chi-siamo/organizzazione/
comitato-interministeriale-per-la-sicurezza-della-repubblica-cisr.html
[13] https://www.commissariatodips.it/profilo/cnaipic/index.html

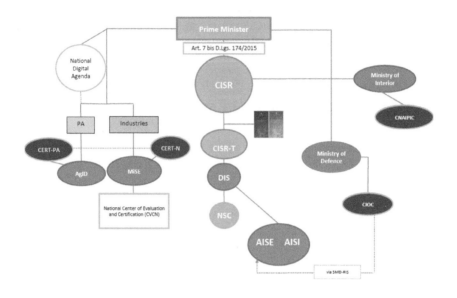

Figure 2.4: Cybersercurity National structure as reported in the White Book on Cybersecurity (2018:18) [147][14]

turn, will represent the *comparable* corpus that is meant to provide a correlation or contrast resource to run the mapping task with the objective of finding equivalents or to check new preferred terms occurrence to be juxtaposed to the ones selected [22]. Firstly, being the scope of the work the Italian context for the Cybersecurity field of knowledge, the existing Italian resources will be presented followed by the English ones. Both of the sets of authoritative documents have been exploited in the terminological processing executions using statistics algorithms to test the coverage level with respect to the outputs in the term extraction results deriving from the processing of the source corpus specifically created for the purposes of the thesaurus' realization, as demonstrated in section 4.2.2.

2.5.1 Italian resources

The Italian framework has been increasingly focusing on Cybersecurity events, both for what concerns the malicious attacks episodes to be analyzed from a informative and legislative perspective and for what regards the guidelines to provide in order to cope with the cyber threats occurring in a range of

---

[14]http://aims.fao.org/standards/agrovoc/concept-scheme

working environments, e.g., the politics (e.g., e-voting), the economics, the academic world, industrial, communication, arts and culture, and many others [147].

A group of guidelines and vocabularies have been published by Italian authorities, and they represent a reference set of documents to be consulted for the creation of other means to monitor the Cybersecurity domain, particularly from a semantic perspective in this specific case.

■ **Glossary of Intelligence – *The language of informative organisms***[15] offers a political glossary containing both domain-specific terminology and regulatory term banks that are more linked to the government environment. This list of terms has been exploited for the purposes of the research work in reaching a threshold coverage score for the Cybersecurity information representation, even though the terms included here are marked by a more generic sign or more governmental perspective oriented, e.g., *functional guarantees*, or *sources managment* are, for instance, terms that have not been included in the more specific Italian Cybersecurity thesaurus because they were not specifically required by the project objectives that saw the need of having a semantic resource reflecting the technical concepts of the domain.

■ **National Framework for Cybersecurity and Data Protection**[16] inspired by the Cybersecurity Framework which has been published by NIST (National Institute of Standards and Technology) is a operational tool for public and private institutions which need a range of techniques to protect personal data, and provides a useful support guide to depict an accurate and domain-oriented cybersecurity strategic plan.

■ **CERT Guidelines**[17]: Computer Emergency Response Team's documents that offer an outline of the strategies that should be employed to overcome cyber threats, among these there are the guidelines for (i) mobile devices security; (ii) ransomware attacks; (iii) behavioural rules for social networking; (iv) cloud computing.

■ **CLUSIT reports**[18]: The Italian Association for Cybersecurity provides an overview of the main cyber crime attacks which have globally occurred in the corresponding years of the publications with respect to a sample of about tens of thousands of serious malicious attacks. At the end of each yearly report, a glossary of the most representative terms is provided and has constituted one of the comparison documents with which the extracted term list from the source corpus and the flat list of

---

[15] https://www.sicurezzanazionale.gov.it/sisr.nsf/archivio-notizie/nuova-edizione-del-glossario-intelligence.html
[16] https://www.cybersecurityframework.it/
[17] https://www.agid.gov.it/it/sicurezza/cert-agid
[18] https://clusit.it/rapporto-clusit/

the thesaurus, which has been approved by the domain experts, has been computationally mapped.

2.5.2 English resources

As for the Italian context, also the English sources from authoritative institutions have been taken into account as gold standards for comparable documents to be mapped with the Italian ones. The following are the official standards, glossaries and vocabularies which have provided reliable materials on which to rely in retrieving trustworthy information about the Cybersecurity domain.

- **NIST 7298r2** – *Glossary of key information terms*[19]: represents the official international standard that provides the glossary of the most representative and technically domain-oriented terms of the Cybersecurity field of knowledge. The glossary included in Nist 7298 together with the taxonomy inside the ISO 27000:2016 standard have been manually translated into Italian language in order to define them as the main gold standard with which the term list extracted from the source corpus and the present thesaurus developed for the OCS project – core element of this work – have been aligned with. The objective of this task is to realize two Italian semantic tools that gather, as much as possible, the specific information about the domain and its relation networks and variations.

- **ISO 27000:2016** – *Information technology – Security techniques – Information security management systems – Overview and vocabulary* [20]: involves a set of key terms referring to the domain of Cybersecurity, alphabetically ordered and with brief description and some cross-reference links representing synonyms and associative terms.

- **NICCS** – *National Initiative for Cybersecurity Careers and Studies – A Glossary of Common Cybersecurity Terminology*[21]: it integrates the terminology contained in the NIST 7298 and aims at providing a range of terms to guide the understanding of the domain together with annotations and definitions.

- **Sophos Threatsaurus** – *The A-Z of computer and data security threats*[22]: offers a list of the main computer and security threats that can be present in informatics and telematics institutions, each one of them is accompanied by the explanation and a brief description on the way threats can take shape within the organizations and software systems.

---

[19]https://www.nist.gov/publications/glossary-key-information-security-terms-2
[20]https://www.iso.org/standard/66435.html
[21]https://niccs.us-cert.gov/about-niccs/cybersecurity-glossary
[22]https://www.ncs-support.co.uk/news/the-sophos-threatsaurus-a-z-of-security- threats/

■ **ISACA** *Cybersecurity Fundamental glossary*[23]: it offers another list of terms related to the domain of Cybersecurity in 17 languages.

For what concerns the main references in the area of ontologies development for the Cybersecurity domain, it is worth mentioning the works addressed to the creation of conceptual models such as [20] [27][110], or the ones focusing on the techniques exploited to set out an architecture for Cybersecurity standards [130] and enterprise's Cybersecurity metrics [23]. Among these studies, for example, Aviad (2015) [6] described how an ontology has been projected towards the data incorporation coming from heterogeneous sources, while missing a common shared terminology, providing a sufficiently complete representation of possible cyber threats, guaranteeing in this way to the institutions the execution of reasoning procedures and the decision-making processes referred to the security sphere. Takahashi (2015) [169] presented a reference ontology for Cybersecurity operational information, which has been implemented, as in this research project, in a collaborative joint work together with Cybersecurity organizations aiming at reviewing industry specifications, defining the different kind of Cybersecurity information with the roles and operation domains. Doynikova [55] (2019) describes the development of an ontology of metrics for Cybersecurity assessment. The baseline relates to the definition of concepts and relations between primary features of initial security data and forming a set of hierarchically interconnected security metrics, where the contexts of application are proved within a case study presentation. What is remarkable in this recent study is how the security metrics are depicted as separate ontological instances, and this lets the concepts relations to be used in order to measure the integral metrics reflecting the security state.

2.6   Group of experts supervision

Vivaldi (2007) [91] underlined the bijective process involving terminologists and domain experts in the way these latter evaluate the semantic resources, created after several text mining steps and scores attributions to reveal the best candidate terms. Indeed, the author affirms:

> "[...] another problem arises in the process of evaluating a TE: who determines what are the relevant terms in a given test text? This issue arises because two different actors with different profiles are involved: a terminologist, expert on deciding whether an expression is a real term or belongs to the general language, and a domain expert, who uses a specific expression to refer to a concept in the domain. This point has often not been taken into consideration." (2007:2)

---

[23]https://www.isaca.org/resources/glossary

As described in the next chapters, the role of domain experts results to be crucial in the insertion of terms in the semantic resources. It is thanks to their skills and expertise that the most difficult interconnections among terms derived from technical documents are possible to be properly structured according to the concrete and specific semantic usage within the domain of interest. For example, the case of the security properties, *integrity, authenticity, availability, confidentiality, privacy, no-repudiability,* are representative for the way they should have been linked to some other key terms, such as *blockchain*, or *Message Authentication Code*, ect., links that have firstly arisen consulting the experts of this field of study.

Terminologists activity should not aside from the approval of the experts [161] because it is thanks to their competencies that a semantic tool can be able to structure and monitor technical terminology and represent, by consequence, a reliable means of control as well as interoperable sources for several organizations that will have to spread official and standardized information on the technical domain of study. Indeed, the services provided by the OCS platform have been designed to reach public and private institutions which might need to interface with a series of cyber risks and threats occurring within the organizations and coming from outside. It is for this purposes that providing a controlled vocabulary and an ontology on the domain of Cybersecurity in Italian, and then in the equivalent English terminology, supervised and approved by experts working within this area of study, could represent a useful way on which various professional groups working in this sector can rely as a reference resource to consult. Moreover, it is possible to exploit in a standardized way the domain-specific terminology thanks to the syntax proper to both thesauri and ontologies when it comes to collaborating on the domain issues to be solved both from a technical point of view and a regulatory one.

To be able to provide a form of normalization of domain terminology, not only a text processing execution is fundamental, as it will be thoroughly presented in Chapter 4 of this book, but also a cooperation with the experts of the domain, who, in accordance with the standards definitions and the updated guidelines of the sectors, can provide an ineluctable seal for the structuring of the semantic framework.

Chapter 3

Related works

3.1 KOS

The *Knowledge Organization Systems* (KOS) constitute a wide range of resources functional to organizing and collecting information belonging to specific areas of study, as affirmed by Hodge (2000:11) [66]:

> "Knowledge organization systems are used to organize materials for the purpose of retrieval and to manage a collection. A KOS serves as a bridge between the user's information need and the material in the collection. With it, the user should be able to identify an object of interest without prior knowledge of its existence. Whether through browsing or direct searching, whether through themes on a Web page or a site search engine, the KOS guides the user through a discovery process."

The author gives a list of the main representative semantic tools that are considered, and addressed in a more extensive way by other experts [127, 152], the best ways in which the terminology of a particular field of knowledge can be managed following a specific purpose. In particular, these are grouped under three main categories: (i) **term lists** (authority files, glossaries, dictionaries and gazetteers), which are characterized by an organization of knowledge that is not highly structured; (ii) **classifications and categories** (subject headings, classification schemes, taxonomies and categorization schemes), which, in turn, present a hierarchically structured knowledge management and aim at creating subject groups; (iii) **relationship lists** (thesauri, semantic networks and ontologies), showing a highly structured system of connections between terms and concepts. In Soergel (2009) [164], there is an expanded set of other KOSs taken into account which the author divides according to the KOS functionalities, e.g., KOS

DOI: 10.1201/9781003281450-3

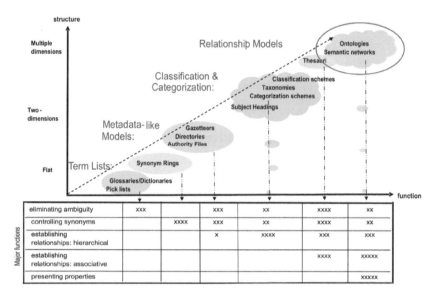

Major functions						
eliminating ambiguity	xxx		xxx	xx	xxxx	xx
controlling synonyms		xxxx	xxx	xx	xxxx	xx
establishing relationships: hierarchical			x	xxxx	xxx	xxx
establishing relationships: associative					xxxx	xxxxx
presenting properties						xxxxx

Figure 3.1: KOS diagram classification [127]

with a focus on concepts and their words, or signs, or focusing on editorials or individuals. Among the main benefits the author underlines there is the possibility to obtain a "road map" to establish connections among the several areas of study placing terms and concepts in exact semantic contexts, then attributing to several concept specific unambiguous terms with definitions to "promote consensus on concepts and terms and the move towards a common language for a field. Provide shared understanding for collaboration, especially computer-supported cooperative work" (2009:8). Another important element stressed by the author and also in the studies carried out by Hjørland (2008) [36], is that KOSs are especially exploited to discover information in big databases, to help users in indexing information and supporting the information retrieval operations. Indeed, they facilitate data organization and modeling as well as metadata structuring and maintenance and interoperability of the information to be shared. As Figure 3.1 shows[1], thesauri and ontologies, which are the two semantic means elected for the purposes of this thesis, are positioned among the highly structured resources to organize information because of their peculiarity in connecting terms by using semantic relationships of hierarchy, synonymy and association and by abstracting the conceptual level of the domain with customized properties that link together the classes of the areas of study.

---

[1] http://aims.fao.org/standards/agrovoc/concept-scheme

3.1.1 Thesauri

As stated in the Standard ISO 25964-1 for *Thesauri and interoperability with other vocabularies* (2011:12) [83], a thesaurus is considered as a

> "Controlled and structured vocabulary in which concepts are represented by terms, organized so that relationships between concepts are made explicit, and preferred terms are accompanied by lead-in entries for synonyms or quasi-synonyms. The purpose of a thesaurus is to guide both the indexer and the searcher to select the same preferred term or combination of preferred terms to represent a given subject. For this reason a thesaurus is optimized for human navigability and terminological coverage of a domain."

The goal in developing a semantic resource such as a thesaurus is to provide to users a list of preferred terms representing in an unambiguous way single concepts proper to specialized domains of study. The network of semantic relationships that characterizes the thesaurus' configuration represents the point from which to start the process of structuring, in an univocal way, the terms that are the representation of specific domain's concepts in a super-/sub-ordinate way (hierarchy) or by equivalence and associative connections. The construction of such a semantic resource should aim at creating a tool that reaches the needs of a particular institutional organization or users' requirements, for this reason it should be electronically navigable and be as precise as possible in the concepts' configuration through selected terms from a range of official document sources that will then be normalized to achieve an approved version of terminology representation by domain experts [171]. For what regards the principle of unique meaning representation of domain concepts, as the ISO 25964:2011 highlights,

> "Concepts and terms are deliberately restricted in scope to selected meanings. Unlike the terms in a dictionary, which are often accompanied by a number of different definitions reflecting common usage, each term in a thesaurus is generally restricted to whichever single meaning serves the needs of a retrieval system most effectively."

Therefore, other meanings cannot be included in the thesaurus' structure for reasons related to information retrieval functionalities: the indexing process needs to rely on fixed standardized and officially accepted terms belonging to a given domain to search accurate types of information. The advantages in using a thesaurus with respect to a taxonomy or a glossary, which are considered in the SKOS's evaluations as semantic resources missing the semantic relations network configuration among terms, are linked to their semantic management of specific information proper to technical fields of knowledge. In detail, the way by which terms structure this linguistic network is linked to the assumption that each term expresses one unique concept. Moreover, the thesaurus is able to guarantee a systematization of the conceptual modeling of the information by means of specific semantic relations interrelating the terms in an unambiguous way, which support the hierarchical, associative and equivalent associations [18, 148]. Also, Soergel (1995:1) in his work underlines the function a thesaurus holds in relating concepts to single-preferred terms, "A thesaurus is a structured collection of concepts and terms for the purpose of improving the retrieval of information. A thesaurus should help the searcher to find good search terms, whether they be descriptors from a controlled vocabulary or the manifold terms needed for a comprehensive free-text search – all the various terms that are used in texts to express the search concept" [163]. A thesaurus is a resource whose construction follows the guidelines shared in the ISO standards 25964-1:2011 and 25964-2:2013 [2, 83]. Before providing essential rules for structuring the information according to the semantic interconnection the preferred terms chosen as entries provide, the standards define the recommendations related to what indexers or terminologists should avoid in the development process of a normalized semantic means for specialized fields of knowledge. Among these suggestions, there are those referred to the non-use of adverbs, or, for what concerns English, adjective in initial positions of the entry terms, nor the definite articles and non-alphabetical characters. It is also suggested to use the plural forms instead of the singular for the insertion of terms in the thesauri. When it comes to analyzing the way a term should be conceived as preferred or non-preferred in the thesaurus, meaning with non-preferred the selection of the synonyms for the main candidate term (preferred), the standards highlight the importance to comply with the final-user needs – which for this project thesis rely on the objectives of creating a support system meant to guide towards the comprehension of the terminology of the Cybersecurity domain –, of concrete applicability and select the ones the group of experts commonly employ to represent specific concepts within their domain-specific environment. Among the other forms of equivalence relationship configuration, the standards underline that if a variant spelling exists for the preferred entry terms, it should be pointed as its synonym (non-preferred term); while, on the contrary, for what regards foreign terms that are considered applicable to the target domain because commonly used, i.e., *loans*, they should be considered as preferred terms. Other forms of synonyms are considered to be:

- terms linguistically different in their origins;

- popular and scientific names;

- deprecated names substituted by new current usages;

- variant names for new rising concepts;

- terms that are generated starting from different cultures that have a common language;

- acronyms, abbreviations or full names;

- slang or jargon usages.

Dealing with multilingual contexts, different degrees in assigning equivalence to terms representing the same concepts are highlighted as follows:

■ Exact equivalence: this might be the best case, as, following the definition contained in the standard [83], "a concept is represented in every language of a multilingual thesaurus, and it is possible to identify preferred terms that are semantically and culturally equivalent. A cross-language equivalence relationship should be established between the corresponding preferred terms." (2011:51)

■ Inexact or near-equivalence: when terms, for reasons related to culture or connotation, show some difference in their extent usages. In this case, the terms are selected, if representing the same concept, as preferred for each language.

■ Partial equivalence: it occurs when a language does not present a term that could represent a broader or more specific concept. The standardized definition is that this case occurs "If the scope of the concept represented by one term falls completely within the scope of the concept represented by the other, the terms are said to be partially equivalent. If the difference in scope is small enough, it might be acceptable to admit the terms to the thesaurus, treating the two as equivalents that both represent the same concept[...]." (2011:51)

In the ISO 25964-1 document there is a focus on the creation of the semantic relations inside the thesaurus among its preferred lead-in terms, i.e., the connections that differentiate the thesaurus from the other means of semantic organization systems since they provide a strong layered outline for the concepts retrieved in technical information textual databases. Specifically, these relations, which in this research project are meant to be automatized through (i) **variation** detected by the automatic terminological extraction, (ii) **word embedding** results in aligning together proximal terms and (iii) **pattern recognition system**, are officially described as below.

Thesauri are realized following the principle of semantic connections among terms that are meant to reply the conceptual modeling of information characterizing a specialized domain by linking them together by three main standardized forms of connections:

1. **Equivalence relation**, is expressed by the tags Used (USE) and Used For (UF), and it covers the synonymy connection between candidate terms: preferred term or descriptor is used to indicate a concept, it is marked with the tag Used (USE), the non-preferred term gathers its synonyms or other meaning associated to the preferred one, it is useful as entry term to access the concept to which is referred and it is marked with the tag Used For (UF). As underlined in the standard, it is important to decide

the criteria on which to rely according to the final-usage needs, meaning that if the thesaurus outline starts by considering as preferred terms only the most technical ones, the rest of lead-in terms should be consistently selected according to the same principle. To better understand the synonymy semantic connection, we provide few examples:

■ Usage:

> *Cyber minacce* UF *Cyber Threat Actors*; *Cyber Threat Actors* USE *Cyber Minacce*

■ Acronyms:

> *Virtual Private Network* UF *VPN*; *VPN* USE *Virtual Private Network*

■ Synonymy control:

> *Cyber attacks* UF *Cibernetic attacks*; *Cibernetic attacks*; USE *Cyber attacks*

2. **Hierarchical relation**, specified by using the tags `Broader Term` (`BT`) and `Narrower Term` (`NT`), is based on the assumption that one term is subordinate as its members or parts to the other one that is the most generic; as stated in the aforementioned standard 25964-2011:

> "The main function of hierarchical relationship is to help both indexers and searchers choose level of specificity." (2011:58)

In detail:
- broader term is a descriptor or a preferred term expressing the concept (top-level term) in a broader way. It is marked with the tag BT;
- narrower term is a descriptor or a preferred term expressing the concept it refers to in a less specific way. It is marked with the tag NT;
A hierarchical relationship exists if there are two concepts and one of them is part, or is included in the other. There are more types of hierarchical subpartitions, that are useful in the perspective of migrating the thesaurus terminology into an ontology structure: the generic relationships tagged with the labels BTG and NTG, partititive connection, i.e., BTP and NTP, and instantiative relation, i.e., BTI and NTI. These latter type of connections are useful when it comes to transpose the semantic structure and hierarchical subdivision of the thesaurus into an ontology outline, as Cardillo *et al.* (2014:5) in their study on the conversion of the thesaurus in an ontology, affirm:

> "[...] thesaurus concepts and facets are treated as classes in the corresponding ontology, while thesaurus terms (synonyms or quasi-synonyms introduced by the equivalence relationships) are represented in the ontology as class labels. In addition, checking and

refinement of thesaurus relationships are performed in order to: explicitly distinguish between different types of hierarchical relationships, manage cycles, orphans, polyhierarchy, etc. (e.g., the *generic relation* (NTG/BTG) are translated into *is-a* relation in the ontology, and the *instantial* one (NTI/BTI) into the distinction between class and individuals)." (2014:5)

"A narrower concept mapped via the BTI/NTI relation generally becomes instance of the OWL class obtained from the broader concept." (2014:8)

■ Whole/parts:

> *Vulnerabilities* NT *Software vulnerabilities*; *Software vulnerabilities* BT *Vulnerabilities*

■ Class/member:

> *Logic bombs* NT *Elk Cloner*;
> *Elk Cloner* BT *Logic bombs*

3. **Associative relation** is marked through the standardized label Related Term (RT) and refers to the coordinated terms associated to other specific to the domain, these coordinations are not related neither to hierarchical structures nor to synonyms configurations, terms with related connection are considered as siblings when having the same broader hierarchy. Because of the weak level of specificity in this type of semantic boundary, a re-engineering process of the thesaurus has been started towards the more explicit one in the representation of the links among domain-specific concepts, i.e., ontology. The related term is a descriptor or a preferred term of a concept that presents an association with another concept with regard to which there is not broader or narrower relationship. An example of RT is:

> ■ *Cyber war* RT *Cyber weapon.*

A useful feature thesauri add in their intrinsic structure is the employment of notes, which are defined under the tags:

■ Scope Note (SN): the term definition retrieved by the source authoritative documents within the corpus

■ History Note (HN): "A note recording changes to a concept or a term." (ISO 25964:2013)

■ Editorial Note: used by thesaurus editors for the modifications

The notes help the searcher in attributing the definition of the concept to the respective representative terms found in source corpus authoritative documents. It proves to be an effective element in assigning properties when it comes to migrating the thesaurus' content into an ontology, and especially, as the next chapter will describe, when the configuration of the patterns has to be run on a corpus to automatize the process in the creation of semantic relationships. An example of SN can be:

> *Phishing* – SN: Tricking individuals into disclosing sensitive personal information through deceptive computer-based means.

Large amount of works in the literature address to the construction of thesauri for a range of different domains of study [22, 45, 62, 65]. One of the works about the methods for building a thesaurus is represented by Broughton's handbook [171]. In this work, the author gives light to the main guidelines to develop a semantic tool through which technical concepts can be organized by means of hierarchical, equivalence and associative relations between the terms representing them in a thesaurus. On the same research extent as Broughton (2006), Soergel (1995) too gives light to the most known thesauri with clear examples of their concrete applicability, such as the AAT thesaurus for Art and Architecture[2]. One of the methods followed to build a thesaurus, beyond the systematic one that uses the main extracted terms obtained by the controlled lists of domain-oriented datasets and structure them in the tangled network of semantic relations, is the **faceted** classification technique that assigns a facet to gather terms with the same connection under a category outlining their structures. It comes from Ranganathan's theory of Colon Classification which categorizes the events of the reality by six main classes/facets under the whole acronym PMEST: Personality, Matter, Energy, Space, Time (PMEST), that in the late 1950s and 1960s have been complemented by the Classification Reserarch Group (CRG) with other categories, i.e., Parts, Materials, Property, Process, Operation, Agent, Space, Time and Form of presentation where facets because these ones are conceived as not *a priori* categories to classify subjects [16], but " [...] derived from the subject to be classified, in each specific context" (2017:288). The methodology of assigning facets to conceptual groups is described in ISO 25964:2-2013: "Firstly subjects are analyzed into simple concepts according to fundamental categories such as activities, entities, places, etc. These concepts are enumerated in the scheme as classes, each with an assigned notation. Then the notation for a complex subject is synthesized by combining the notation of the simpler concepts in accordance with rules for sequence, known as citation order. The sequencing rules are essential to ensure that all the documents on the same complex subject are given exactly the same notation and hence are brought together." (2013:52) This principle is adapted to thesauri construction, such as the aforementioned AAT, which exploits the grouping system provided by facets to organize knowledge by upper categories and better segment the content of the domains to be managed.

---

[2]https://www.getty.edu/research/tools/vocabularies/aat/

3.1.2 Ontologies

As stated in Staab and Guarino (2009:3) [137]:

> "Computational ontologies are a means to formally model the structure of a system, i.e., the relevant entities and relations that emerge from its observation, and which are useful to our purposes. [...] The ontology engineer analyzes relevant entities and organizes them into concepts and relations, being represented, respectively, by unary and binary predicates. The backbone of an ontology consists of a generalization/specialization hierarchy of concepts, i.e., a taxonomy."

And according to Gruber (1994) [170] the ontology is an "explicit specification of conceptualization." Indeed, the ontology provides a conceptualization of the objects that belong to specific domains of study managing these latter in a structure of concepts, or better said, a range of constructs which describe the world by displaying classes, individuals and properties, which are the abstract representations of the domain's reality and are organized through structured connections that serve to infer information of the field of knowledge [45]. The underlining logic applied to construct the ontology is most commonly a formalization of the higher first-order logic, i.e., Description Logic (DL) ontologies [59]. From a practical point of view, the methods followed for building ontologies observe basic principles that can be found in guidelines such as the one published by Noy and Mcguinness [140] or Bourigault [45]. There are several ways by which ontologies can be constructed:

■ the manual development makes use of NLP operations on texts especially referred to post-processing and clustering operations; as Hernandez (2005) deeply analyzes, terminolgists can rely on a set of pre-establibiled methodologies, as OntoClean (Guarino 2002) or OntoSpec (Kassel 2002), and provides a list of platforms that can guide the creation of ontologies as Protégé – that is the one used for the creation of the Italian thesaurus for the Cybersecurity domain; ODE (Ontology Design Environment); OntoEdit; WebOnto, ect.

■ the computer assisted methodology, which consists in populating the ontologies by exploiting existing tools that help in organizing the classes in a tangle of hierarchical structure as *is-a, is-kind-of.* Nazarenko (2008) [167] provides an evaluation of the main systems for conceptualizing the terms extracted from a corpus into an ontology configuration; the author, in fact, provides a list of software to use to support users in building ontologies starting from given textual datasets:

1. **Text2Onto**, which has as benefits:

> "[...]two new paradigms for ontology learning: (i) Probabilistic Ontology Models (POMs) which represent the

results of the system by attaching a probability to them and (ii) data-driven change discovery which is responsible for detecting changes in the corpus, calculating POM deltas with respect to the changes and accordingly modifying the POM without recalculating it for the whole document collection." (Cimiano, 2005:228)

2. **Terminae**, is a system that works with term extraction, concordances, synonymy retrieval meant to be key elements for the ontology's realization [135].

3. **OntoGen**, provides a semi-automatic methodology for ontology construction by proposing to the experts some concepts in the form of documents classes by exploiting non-supervised and supervised algorithms, e.g., *K-means*, *SVM active learning* [33]. As Nazarenko (2008:4) underlines, OntoGen is a semi-automatic approach of conceptualization in which the classifier runs on the collections which are associated to main concepts.

Being written with the languages of the Web Semantics, RDF and OWL, ontologies provide a relevant advantage of interoperability between informative systems, a feature thesauri has in a restricted way through their exportation in XML. OWL employs RDF powerful syntax to represent the triples `subject-predicate-object` through graphs, and applies high sophisticated reasoning engines to start inferences on the classes created in the ontology structure which enable developers to establish the ranges of classification, equivalence and subsumption, that, in turn, can lead to the ontology inconsistency recognition [157].

Similarities between ontology can be identified in the way both of them describe and organize a domain, include concepts and relations between them, use hierarchies, and use terms to represent concepts. Both can be applied for information management for cataloguing and in search engines. With respect to the discrepancies, first of all, thesauri have as their main purpose that of being used in librarian contexts as indexing tools and controlled vocabularies, therefore, they are meant to represent knowledge in a less formal and comprehensive way with respect to ontologies. On the other hand, ontologies are characterized by a high level of conceptual abstraction as well as formalisms in describing domain knowledge. Ontologies are structured in a way that the explicit representation of the types of relationships is provided alongside the employment of efficient formalisms, which are not possible to define within thesauri (e.g., axioms, relationships, cardinality). Thus, to represent hierarchical relations between classes and subclasses two declared relations *is-a* and *kind-of* are used, and for the whole-part relation which represents the meronymy between classes, the transitive property *partOf* relation is employed. As seen before, in the thesaurus those relationships are treated as hierarchical relationships. Finally, the associative relations in an ontology is made explicit according to the exact connection (predicate) that exists between two classes.

The interoperability of semantic resources like thesauri and ontologies, guided, among others, by the principles of linked open data [121, 162], guarantees a shareable knowledge organization system that can facilitate the coordination among several users for different terminological tasks. Having the prospect of creating a language which aims at allowing a sound structure of interaction among informative systems, the ontology, with the use of RDF and OWL languages, guarantees a higher level of re-usability in the information provided [53, 137].

Ontologies, as summarized by the W3C Consortium[3], are characterized by a set of fixed elements that outline their structure which can be applied to several specific domains:

- **Classes**: is a group of lexical forms organized in taxonimic way. There are *disjoint classes* when an axiom `DisjointClasses(C1 ... Cₙ)` expressing that all the individuals cannot be contemporary an instance of both classes $Ci ... C_j$ for $i \neq j$.

- **Axioms** are assumptions related to concepts or relations of an ontology considered to hold True values. In OWL syntax axioms can be provided through declarations, they can regard object or data properties, datatype, keys, assertions or annotations. An example given by W3C Consortium: SubClassOf(Annotation(rdfs:comment "Male people are people.") a:Man a:Person).

- **Subsumption**: is the relation referring to the inference of the first concept from the second one through the integration of a property the first one bears; the second concept generalizes the first one, i.e., $<<$ generalization/specialization $>>$ of two concepts.

- **Individuals**: the domain's objects, i.e., the *named entities* that can represent the explicit names corresponding to the same objects.

- **Literals**: respective to RDF concepts are characterized by strings, labeled as IRIdatatypeIRI or integers.

- **Attributes** are the values which objects and class present.

- **Datatypes**: are the entities equivalent to the classes apart from the fact that datatypes include individuals while classes strings and numerical values.

- **Object properties**: are relations identified with the IRI owl:ObjectProperty that correlates pairs of individuals.

---

[3]https://www.w3.org/TR/owl-ref/

■ **Data properties**: are relations that connect individuals with literals, identified with the IRI owl:DataProperty.

■ **Restrictions**: are statements that point out the true value of assertions in order to be conceived as input, the main ones can be: cardinality restriction, data properties restriction.

■ **Annotation semantic**: is given by correlating concepts to a specific object.

Ontology typologies

There are three OWL sublanguages that developers can choose in creating domain specific ontologies with respect to the formalism levels meant to be reached. W3C recommendation[4] reports the specific features for each one of them that can be summarized as follows:

1. The first sublanguage is called **OWL Lite** and is the one that supports the migration into thesauri and taxonomies. It provides cardinality constraints with only 0 or 1 values.

2. The second is **OWL DL**, where DL stands for *Description Logics* the logic of the formal foundation of OWL. It guarantees a formalization through the usage of restriction and enables reasoning as well as inference. This is the one that has been created for this thesis research project on Cybersecurity knowledge management and representation.

3. The third sublanguage, that is the most expressive, is the **OWL Full**. "OWL Full allows an ontology to augment the meaning of the pre-defined (RDF or OWL) vocabulary."[5]

3.2   Semantic conversions – from thesauri to ontologies

The major constraint thesauri show is the flat visualization of the associative semantic connections that can be a limitation in the representation of the conceptual structure of the domain information [54]. Indeed, the ontology is characterized by an higher level of abstraction that grants a detailed conceptualization of the specific knowledge, whilst thesauri are more commonly represented by indexing features that fix as its objectives those of helping searchers and indexers in organizing the information of technical fields of knowledge by using pre-established and standardized connectors to which terminologists must comply with. As specified in ISO 25964-2013 [90], ontologies give the chance of creating more explicit relations between concepts, this is possible by using the formalisms (cardinality, axioms) and properties referring to the concepts and their connections.

---

[4]https://www.w3.org/TR/2004/REC-owl-features-20040210/#s3
[5]*Ibidem*

Another reason why thesauri are often converted into an ontology system, for example MeSH or AGROVOC [114], is because the explicit grammar behind the ontologies is the SKOS/RDF which enables to share the information created through the semantic resources [122], granting in this way a better form of inter-operability with respect to the one offered by a thesaurus with its standardized labels and fixity. Among the methods mostly employed to migrate the content inserted in the thesaurus towards an OWL systematization proper to ontologies, there are the techniques that start from the thesauri standard structures, usually configured according to the ISO 25964-1 and 25964-2 criteria, choose the pre-ferred terms of the thesauri and assign them the RDF labels with unique identi-fiers, which are going to be then interrelated with other properties to clarify the synonymy or hierarchical connections [123]. Kless (2012) [49] explains how the thesaurus can be converted in an ontology by enhancing its semantics with OWL rules: for instance, each concept becomes a class in the ontology, the hierarchi-cal connections relation explicated in the ontology by the link *is-a* or *is-part-of* standing for the super-ordination level. The author also suggests to take into ac-count the possibility to rely on pre-defined relationships-sets configurations, as for example the one included in DOLCE ontology, that can be cross-mapped with the one that has to be built up. In Wielinga (2001) [39] thesauri are also indicated as valuable resources in the indexing and retrieving processes in the way they provide an organization of domain specific knowledge by putting together pre-ferred terms selected after terms processing tasks and assigning to them specific hierarchical, equivalent and associative connections to other terms in the same domain framework; for this reason, and mainly because of the structured form of information thesauri provide, they represent a valid source from which to start to populate the ontologies following RDF's inter-properties extent.

This study focuses on the need to convert the information included in the Italian thesaurus for the Cybersecurity domain because in some circumstances the standard relationships normalized by the ISO terminology standards (25964-1:2011, 25964-2:2013) proved not to be completely satisfactory in terms of knowledge representation for the technical concepts according to the specific needs of the final-users. The OWL language proper to ontologies can guarantee a higher level of customization in the selection of the links that correlate several terms and allow the informative systems to easily exchange structured informa-tion in a common shared export format to be exploited. The thesauri's developed modeling structure will help in shaping the ontology system for the domain con-ceptualization by taking advantage of its internal disposition of relationships and definitions of the terms that will be essential to create new more fine-grained con-nections in knowledge management of the Cybersecurity field. Indeed, the way OWL-DL is written enables users to have an automated inferring reasoning upon the concepts uploaded that are characterized by the properties of memberships and unambiguousness (Kless, 2012).

3.3   SKOSs systems

Managing technical terms proper to specialized languages represents one of the main tasks of Knowledge Organization Systems (KOSs). Simple Knowledge Organization System (SKOS), as stated by the W3C Recommendation of 2009[6] constitute:

> "[...] a common data model for sharing and linking knowledge organization systems via the Web.
>
> Many knowledge organization systems, such as thesauri, taxonomies, classification schemes and subject heading systems, share a similar structure, and are used in similar applications. SKOS captures much of this similarity and makes it explicit, to enable data and technology sharing across diverse applications."

The language SKOS systems use are RDF and OWL because of their exchangeable nature that make them readable and exploitable from several informative contexts. The model built by SKOS is based on the concept-scheme configuration where a range of concepts are identified by unambiguous URIs `skos:Concept` and are connected by a set of semantic associations and hierarchies. SKOS systems data model, by employing OWL language and RDF graph syntax, are able to merge the abstract configuration proper to ontology settings, which are based on axioms, with the thesaurus and classifications schemes outlines characterized by an organization of concepts through relationships associations. Indeed, each concept is defined by a label `skos-label` or alternative labels, that in most of the cases are synonyms expressed by RDF literals, then SKOS provide several descriptive labels, i.e., lexical or documentation properties [181]. In the context of KOSs, semantic resources, such as thesauri and ontologies, are useful tools to organize domain specific knowledge and to support processes like document indexing, information searching and retrieval and, in some cases, automatic reasoning (e.g., for decision making), above all in those specialized domains where semantic ambiguity between terms represent a step to be avoided.

3.4   Research approaches

In this section, the methodologies exploited for the purposes of this research activity are presented. For what concerns the categorization in text mining extraction process from source textual databases, (i) Latent Dirichlet Algorithm (LDA) has been applied to the set of documents related to the Cybesecurity documents to obtain significant categories to thematically gather the range of data in the source corpus; for the main keywords extraction; (ii) the PKE [40] library has

---

[6]https://www.w3.org/TR/skos-reference/

also been used, which implements a series of keyphrase extraction approaches, e.g., TopicRank, MultipartiteRank, TD*IDF and TopicalPageRank; (ii) regarding the recognition of semantic relations in the field of distributional semantics; (iv) variation (Daille, 2017) in the domain-specific terminology has represented the starting point from which to detect synonyms and hierarchical structures; (v) more sophisticated computer-based algorithms of word embedding, i.e., Word2vec and FastText, have been applied to textual documents in order to retrieve the similarities in terms and levels of proximity that could help in building the automatic identification of the semantic network structure of the Italian Cybersecurity thesaurus; (v) finally, a pattern-based configuration both assisted by pre-trained computer software and outlined by coding has been implemented on the documents used to extract candidate terms with the objective of collecting the most representative semantic recursive occurrences, verbal and nominal, that could match semantically meaningful chains to be conceived as triggers to the semantic relationships construction to be imported in the thesaurus.

3.4.1 Clustering approaches

Part of the activities carried out in this thesis project research has been addressed to the creation of main entries categories in order to manage in a highly structured way first the hierarchical relationships, and secondly the variation in terms of synonyms and neologisms. In early Information Retrieval (IR) studies, gathering documents was considered the main technique to guide text mining tasks, as Yaari (1997) [180] and Lin (1998) [107] point out in their works; going further in recent surveys there are ones referring to the topic modeling techniques [144] that use computing models to be trained in order to categorize large textual datasets. For whar regards the Latent Dirichlet Algorithm (LDA) [74] following Blei's (2003:2) statements,

> "In LDA, we assume that there are k underlying latent topics according to which documents are generated, and that each topic is represented as a multinomial distribution over the IVI words in the vocabulary. A document is generated by sampling a mixture of these topics and then sampling words from that mixture. More precisely, a document of N words w = (W1,''' ,W N) is generated by the following process. First, B is sampled from a Dirichlet(a1,''' ,ak) distribution. This means that B lies in the (k -I)-dimensional simplex: Bi 2': 0, 2:i Bi = 1. Then, for each of the N words, a topic Zn E I , ... ,k is sampled from a Mult(B) distribution p(zn = iIB) = Bi. Finally, each word Wn is sampled, conditioned on the znth topic, from the multinomial distribution p(wl zn). Intuitively, Bi can be thought of as the degree to which topic i is referred to in the document."

To sum up, the distribution of terms and words characterize each documents that is made of a group of topics, the model is able to detect the likelihood a document's terms distribution has to belong to given topics. Still in Blei, topics are

> "[...] represented explicitly via a multinomial variable Zn that is repeatedly selected, once for each word, in a given document. In this sense, the model generates an allocation of the words in a document to topics. When computing the probability of a new document, this unknown allocation induces a mixture distribution across the words in the vocabulary. There is a many-to-many relationship between topics and words as well as a many-to-many relationship between documents and topics." (2003:8)

3.4.2 Keyphrases extraction

Automatic keyphrases extraction is a useful process used to index or retrieve documents from big data or large-scale corpora consisting in extracting the main keyphrases from a given document [177], they can be constituted by single or multi-word units of lexicon (Daille, 2017). Hasan (2014) [104] provides a list of the existing methods to automatically extract keyphrases from source datasets, both unsupervised and supervised, highlighting also the "corpus-related factors" that influence the correctness of keyphrases' extraction execution. In detail, these are: the length of documents that impact on the identification of candidate keyphrases; the structural consistency, meaning that some of the keyphrases are most frequently retrieved in certain portions of the text to be analyzed; topic change, that is the talking points that are different in time as the keyphrases associated to topics; topic correlation, with this the authors mean the association one keyphrase has with another one (2014:1263). The keyphrase extraction library chosen for the purpose of this thesis project is PKE (Boudin, 2016). PKE is an open-source python keyphrase extraction toolkit that implements several keyphrase extraction approaches and enable users to get a semi-automatic organization of terminology that could represent the hierarchical and associative semantic chains, otherwise normalized just with the manual evaluation of terminologists in assigning standardized tags, i.e., BT/NT and RT. The methodology used in extracting keyphrases from the source corpus, which has been pre-processed in order not to create noise and ambiguity in the terminology provided [60], has been oriented towards unsupervised approaches, included as PKE models, specifically to MultiPartite Rank, TopicRank, and TF*IDF measures. As stated in Gallina (2020):

> "MPRank, relies on a multipartitegraph representation to enforce topical diversity while ranking keyphrase candidates. It includes a mechanism to incorporate keyphrase selection preferences in order to introduce a bias towards candidates occurring first in the document. MultipartiteRank was shown to consistently outperform other unsupervised graph-based ranking models. TF×IDF makes use of the statistics collected from unlabelled

data to weight keyphrase candidates. As such, it often gives better results."
(2020:4)

Topic rank model, following the words of Bougouin (2013) [3]:

> "is an unsupervised method that aims to extract keyphrases from the most
> important topics of a document. Topics are defined as clusters of similar
> keyphrase candidates. Extracting keyphrases from a document consists in
> the following steps [...]. First, the document is preprocessed (sentence seg-
> mentation, word tokenization and Part-of-Speech tagging) and keyphrase
> candidates are clustered into topics. Then, topics are ranked according
> to their importance in the document and keyphrases are extracted by se-
> lecting one keyphrase candidate for each of the most important topics."
> (2013:544)

> "TopicRank represents a document by a complete graph in which topics
> are vertices and edges are weighted according to the strength of the se-
> mantic relations between vertices. Then, TextRank's graph-based ranking
> model is used to assign a significance score to each topic." (2013:546)

The extraction of the best keyphrase candidates *per* topics provides a useful
trigger to the categorization of the information about the specialized domain of
Cybersecurity, belonging to particular topics gives light to the semantic subordinate relationships terms share with the main head-terms [48]. One important
measure to increase the precision in the results obtained with keyphrase extraction as well as term extraction processing is the term frequency and inverse document frequency measurement, i.e., TF*IDF computation. As Gaudivada (2018)
[73], specifies:

> "Representing documents and queries using term frequency helps to
> achieve only one of the indexing goals—recall. The term frequency mea-
> sure does not capture the importance of terms that occur rarely across indi-
> vidual documents of a collection. Such rare terms are useful in distinguish-
> ing documents in which they occur from those in which they do not occur.
> Inverse Document Frequency (IDF) measure is an appropriate indicator of
> a rare term as a document discriminator. The idf measure helps to improve
> precision. This suggests that both the term frequency and the inverse doc-
> ument frequency measures can be combined into a single frequency-based
> measure using the tf-idf-based term weighting to improve both precision
> and recall." (2018:331-341)

Also, Nguyen (2014) [141] explicates the formula of inverse document measure as:

"

$$idf_i = log \frac{|D|}{d:t_i \in d} |d$$

where |D| is the number of documents in our corpus, and |$d : t_i d$| is the number of documents in which the term appears. If the term ti appears in every document of the corpus, idf_i is equal to 0. The fewer documents the term ti appears in, the higher the idf_i value."The measure called term frequency-inverse document frequency (tf-idf) is defined as $tf_{i,j} * idf_i$ (Salton and McGill, 1986). It is a measure of importance of a term ti in a given document d_j. It is a term frequency measure which gives a larger weight to terms which are less common in the corpus. The importance of very frequent terms will then be lowered, which could be a desirable feature." (2014:95-115)

3.4.3 NLP approaches for building semantic structures

In this section the process of automatically identifying the semantic relations within the thesaurus is presented by describing the selected distributional similarities methodologies. Firstly, the variants recognition procedure is presented, successively the application of word embedding models is described together with the analysis of the task concerning the retrieval of recursive fixed expressions, i.e., patterns, inside the documents that make up the source corpus, by exploiting the natural language's phrasal structures in order to set out a methodology of semi-automatic identification of specific semantic connections among domain-oriented terms will be described, and, in the end, a way to detect new terminology within the results.

3.4.3.1 Distributional similarity

This section provides a delineation of the state-of-the-art approaches concerning the detection of variants in given terminology lists as a result of the semantic processing tasks, that will help through the hierarchical configuration of terminology in the Cybersecurity domain, as well as in the identification of synonyms, the word embedding models that help out in discovering the proximity level independently from languages among terms in a given source corpus, and a pattern-based approach configuration.

Variants

As Daille (2005:29) states, a variant of a term is "an utterance which is semantically and conceptually related to an original term," where utterance stands for an attested lexical form that is present in a text. Still in Daille (2005), it is stressed that semantic variants of terms can be considered relevant to start building hierarchical, associative and synonyms relations in the terminology proper to specific domains, specifically with respect to some of the syntactic and morphological clues that help in identifying the conceptual connections between the base-terms. In Dury (2010) [143], morphological variants are triggers from which to start thinking that terms have changed and there could be a "lexical death," i.e., obsolescence that can lead to a substitution of previous terms. De Cea (2012) [72] sorts out a list of three forms of variants referring to original terms: the ones occurred by synonym association, others that represent the semantic distance from the related original terms, and the third form concerns the reference to another term that resulted connected to the officially recognized term through a conceptual bond; subsequently the author highlights the causal factors for the variation in terminology by referring to Freixa (2006) [86] when the author explicates that the reasons why terms can be altered by their semantic variation could be retrieved in the different forms of dialect authors use, or register, as well as distinct discoursive needs, interlinguistic differences and various cognitive conceptualization. Cabré (2008) [115] as well provides some differentiation in the typologies of variants, specifically for the denominative ones, distinguishing the one referred to graphic changes, morphological and linguistic. In Daille (2005, 2017) there is a broader analysis of the semantic variations that can change the conceptual meaning of a term and impact the structure of the semantic resources that reflect a particular field of knowledge. The author identifies specific typologies of variants in lexicons: terms that change for **graphic** modifications, **inflectional** differentiation, **morphosyntactic** variations and types of variants defined **"shallow-syntactic"** and **"syntactic"** meaning that is respectively a variant that modifies the role a word has in the base-term structure and the one that modifies the inner outline of the base-term. The important element that arises in the author survey on variation in terminology, is that through a range of relational adjectives it is possible to create the conceptual and linguistic expansion or reduction of given multi-words units of text, granting variants recognition that can lead to the automatic structuring of distributional synonymous forms. This and the syntactic-morphological evidences, such as the indicators for the antonym relation *non*, the suffixes for the actor's role in sentences, such as, for the English language *-er, -ers, -or, -ors*, or the affixes to point out the time category, still in English *re-, pre-, post-*, allow to create hyperonym structures of language between variants and the base-terms, as well as coordinations towards synonyms detection.

Word embedding

Firth (1957) stated that "You shall know a word by the company it keeps," and that coincides with the final achievement word embedding models reach in their application, i.e., the identification of the proximity level between terms occurring in similar contexts in order to retrieve semantic co-occurrences which can lead to group candidate terms to be included in semantic resources and realize fine-grained relationships networks based on models' results found in a specialized corpora. Terminology disambiguation and checking for similarity, as well as the inference in lexicology of recursive structures [142, 154, 166], are some of the results achieved by word embedding models. This means that in a sentence the embedded words converted into weighted numeric-value vectors are inclined to become similar, or sharing some features as the same class association, e.g., *king* and *man* will be strictly close as well as *queen* and *female*, if they appear in a same distributional context [105].

As Iacobacci (2016) [168] points out, citing one of the pioneers of word embedding [176]:

> "An embedding is a representation of a topological object, such as a manifold, graph, or field, in a certain space in such a way that its connectivity or algebraic properties are preserved (Insall et al., 2015). Presented originally by Bengio et al. (2003), word embeddings aim at representing, i.e., embedding, the ideal semantic space of words in a real-valued continuous vector space. In contrast to traditional distributional techniques, such as Latent Semantic Analysis (Landauer and Dutnais, 1997, LSA) and Latent Dirichlet Allocation (Blei et al., 2003, LDA), Bengio et al. (2003) designed a feed-forward neural network capable of predicting a word given the words preceding (i.e., leading up to) that word. Collobert and Weston (2008) presented a much deeper model consisting of several layers for feature extraction, with the objective of building a general architecture for NLP tasks. A major breakthrough occurred when Mikolov et al. (2013) put forward an efficient algorithm for training embeddings, known as Word2vec." (2016:897)

The methodology followed in this thesis is the *Full-compositional word embeddings*, which considers a sentence as represented by an element-wise sum of the word embeddings of semantically related words of its parts (Arora, 2017). The full-compositionality word embeddings models prove to be a valid technique when it comes to retrieving semantic relations, specifically for the automatic extraction of hyperonyms, synonyms [17], related and causative terms. The models applied for the purposes of this project are essentially two: Word2Vec [178], and FastText [26], as section 4.3.4 will describe in detail.

Pattern-based approaches

One of the objectives pursued in this thesis is to find out a method to infer the semantic relations network between the candidate terms of the Cybersecurity thesaurus and detect new forms of terminology inside the domain to guarantee a semantic coverage with respect to the field of study. The pattern-based approach [80] has been tested after the more fine-grained computer-assisted algorithms of keyphrases extraction systems and word embedding models, as well as subsequently to the variant recognition through the use of semi-automatic terminological software manipulation. The goal is to establish a method to discover, through certain frequent semantic chains occurrences – such as causative verbs

– the semantic associations to be reproduced inside the thesaurus and the ontology for the Cybersecurity domain as semantic links. Condamines (2008) [12] considers genre as an impact factor in establishing the accuracy patterns have in the specific domains. For instance, the author discusses the regularities concerning the lexical variation in given domains that are easier to retrieve if the field of knowledge is characterized by a range of documents that employ a technical terminology with highly recurrent semantic structures to be detected. Indeed, the employment of these recursive pattern-based configurations run over a given specialized corpus can help in realizing a structured network of semantic connections in a semi-automatic way, as the hierarchical and causal ones (Roesiger, 2016). The meronymy relation has been thoroughly described in Barriére (2002) [41] as well as the casual relation in texts. Still in Barriére (2008) depiction of the main state-of-the-art works on patterns, the author defines the pattern as relation that involves two terms and a linguistic item that stands for the semantic relation:

> "In its most basic form, a pattern-based semantic relation would include a term X, a term Y, and a linguistic unit expressing a semantic relation between term X and Y. Finding instances of a semantic relation in texts using linguistic patterns can be implemented in different ways. It can be achieved by building a query where both X and Y are unknown terms linked by a known relation, as for example, is-a(X,Y). Another strategy can be applied to retrieve one unknown parameter and set the second parameter to a known value, for example the pattern is-a (X,drug)." (2002:9)

Many authors in the literature refers to Hearst (1992) [126] when the topic addressed is related to the automatic discovery of patterns in domain-oriented datasets, especially with respect to the author's reflections on the hyponymy detection from large corpora by using wildcard characters in comparison with the information found in WordNet concepts and synsets structures[7]. More specifically, Hearst considers hyponym a lexical unit that is part of another concept represented by another lexical item in the native language contexts, and provides some pre-established patterns, that with the use of some adverbial and prepositional phrases, accounts for the membership relation:
- NP_0 such as NP_1, NP_2 . . . , (and|or) NP_n, – such NP us NP ,* (or| and) NP;
- NP, NP* or other NP;
- NP, NP* , and other NP;
- NP, including NP, * or|and NP;
- NP, especially NP, * or | and NP.

Meyer (2001) [79] too gave light on the properties towards the generation of different lexical patterns, specifying that there are a series of *knowledge patterns* that can orientate the recognition of established relations between terms. As the author affirms, these semantic fixed indicators are recurrent expression that are constituted by:

---

[7]https://wordnet.princeton.edu/

"linguistic and paralinguistic elements that follow a certain syntactic order, and that permit to extract some conclusions about the meaning they express." (2001:237)

The comparison methodology employed with Wordnet pre-established hierarchies has also been adopted by Giriju (2006) [149] for discovering part-whole relationships in given textual corpora. The rules the authors apply rely on "iterative semantic specializations (ISS)" techniques that have been adopted on Noun Phrases (NPs); the scheme the author follows to identify the recurrent patterns in an automatic way is a decision tree algorithm having as training set "+ positive" to define the property of inclusion (meronymy) and "-negative" to define the terms that are not characterized by the meronymy relation. This algorithm is also integrated by a semantic condition that the noun constituents "matched by patterns must satisfy in order to exhibit a part-whole relation" (2006:84).

The part-whole relation is together with the causal one the connection meant to be retrieved by using pattern bank structures, which have used lexico-syntactic indicators, adapted to the domain of study terminology, to retrieve the desired output such as causative verbs or prepositions or co-occurrence of multi-word units for the meronymy inclusions. The aim in using patterns configurations related to the causative relations [109] is that of providing an improvement in the structure of the related terms in the thesaurus. In ISO Standard 25964-2 of 2013, when it comes to discussing about the interoperability of the systems, the associative mapping is described as a connection that "[...] may be established between concepts when they do not qualify for equivalence or hierarchical mappings, but are semantically associated to such an extent that documents indexed with the one are likely to be relevant in a search for the other." As can be further observed, the associative relationship in thesauri systematization is among the others, hierarchical and equivalence, the one that presents more ambiguity in the way it connects the domain-oriented terms. By using causative-based patterns the references from one specific term to another seem more precise and solid.

Lefeuvre (2017) [108] broadens the analysis of the relationships related to the hyponomy, meronymy and causal references to be retrieved in large-scale textual source datasets, characterized by a mixed nature of documents typologies (ranging from the scientific to vulgar ones), providing a set of several recursive mark-up pattern banks to be run on the documents, with a clear distinction between those that are frequent in scientific oriented documents and others that appear more often in popular datasets mostly composed by common semantic representations in lexicon, in order to gather all the occurrences of terms referring to part-whole associations and causal relations. For instance, the author deals with the hyponymy-meronymy configurations by associating a range of semantic indicators that occur many times in the texts given in input. To give some examples, the structures of patterns run over the scientific and vulgar corpus have reflected the (i) *inclusion relations*, which denote the hierarchical connection between terms in a given domain, e.g., "X is a type of A. Or X, Y and Z are types of

A. Or A has the specific concepts X, Y and Z. Or A has the subtype X" (2017:21); (ii) members to same genres with the configuration supported by the main part-whole representative verbs, such as *collect, reassemble, collect, group by*, followed by a preposition and then by a DET X DET Y (the order of this structure may change in the patterns retrieval process across the source documents); (iii) junction or fusion structures, which are also arranged through a set of pre-fixed expressions like *joint with, linked to, assembled with, to form, being embedded in*. The other syntactic or nominal constructions serve to the author to identify the cause-effect expressions in the text – e.g., *to generate DET X, creation of Y by DET X* – that could help in the creation of morpho-syntactic relationships for the domain-oriented semantic structure configuration, as it will be better described in the next section 4.3.3 concerning the pattern-based methodology to enhance the associative connection outlined in the thesaurus to be migrated into OWL language.

Chapter 4

Research methodology

In this chapter the methodology followed for the construction of the Italian thesaurus for the Cybersecurity domain will be described, as well as that referring to the automatization of the hierarchical structure identification as well as the synonymy and associative systematization automatic recognition. The first section covers the description of the source corpus compilation that gathers the official information about the domain collecting both authoritative legislative documents and sector-oriented magazines, firstly in Italian language and subsequently in English to realize a multilingual thesaurus for the Cybersecurity field of knowledge. The next sections are addressed to (i) software-aided terminological extractors presentation; (ii) description of the approaches pursued for the automatic recognition of the semantic relationships proper to thesauri in order to facilitate its computer-aided structuring.

4.1 Corpus construction

This section covers the steps carried out towards the compilation of the source corpus that has represented the starting point from which to begin the textual processing to extract the main representative candidate terms for the domain of Cybersecurity in Italian language, foundation of the thesaurus construction. Since the project activity has also been oriented to the multilingual alignment in English, the corpus described herein is supplemented with the target second language comparable textual documents. The corpus for the domain of Cybersecurity includes both English and Italian documents given the mixed language nature of the information about the domain of study. Hence, the realization of the

DOI: 10.1201/9781003281450-4

thesaurus refers to English textual datasets dealing with the same issues as the Italian corpus and that could provide important loanwords which belong to the domain to be analyzed. The perspective of taking into account English sources has as its end the conceptual analysis of the relationships between terms in the English language and the verification of the meanings that terms assume in different geographic contexts and the possibility of matching or not the semantic bonds from a source language to a target one. To obtain a thesaurus which could be the maximum representation of the knowledge level of the domain under analysis, where the matching is not possible, a translation for the English terms selected for the population of the Italian thesaurus is proposed in order to guarantee a representativeness threshold as uniform as possible. Translating the English terms for the domain of Cybesercurity has represented a preferred task to carry out in the multilingual alignment perspective since the application of the same methodology herein described for the Italian documentation would have implied highly considerable efforts in the process of research methods validation.

4.1.1 Documents selection

The documents selected to be part of the source corpus meant to become the input for the linguistic pre-/post-processing tasks, have been chosen in compliance with the principles of the source laws [71] that systematize the order from which to begin the information retrieval referring to specific domains and populate the corpus to be processed. The main starting assumption in collecting the documents, being part of the source corpus, has been grounded on the principle of authoritativeness in order to create a semantic tool as reliable as possible.

The nature of documents has been, for instance, permeated by the (i) trustworthiness of publishers, (ii) their adherence to the domain's issues, (iii) compliance with time range, especially for what regards the legislative dataset which had to contain only normative documents in force, as table 4.1 shows; the official documents status recognized by domain experts, politics associations. Working with a domain under development, the goal to reach is to keep up with the terminological evolution of a dynamic corpus as to test through time the structure of the semantic relationships inside the thesaurus with the upcoming of new terms and with the updating of the existing ones in order to avoid the semantic obsolescence that a continuous proliferation of information would bring forward. Means that are usually employed for information retrieval have been consulted, such as specialized articles, sector-oriented magazines, the latest updates on the subject from the main national newspapers (referring to the episodes of cybercrime, hacker attacks, as "Wannacry", especially to rely on official documents shared within the community of experts to better systematize the knowledge-domain to be represented by the thesaurus); national, regional and European legislation, UNI/ISO standards. In order to retrieve official sector-oriented scientific journals other portals have been selected to start the search, e.g., the National Library services, websites of publishers or institutions that work in the sector, Urlich's Web[1], JStor[2], OCLC WorldCat database[3], Karlsruhe Institute of Technologies database[4], OPAC[5] or BNCF[6], *Registro degli operatori di comunicazione* – ROC register (AGCOM)[7]. More specifically, the definition of the magazines and scientific Cybersecurity-oriented journals corpus implied an upstream search carried out by using the terms contained in the Italian gold standard for the domain of interest, i.e., the Glossary of Intelligence, as keywords to be used in filtering out, among the magazines which are officially registered in the Italian authoritative list, namely the ROC, the titles of issues which most matched with the knowledge-domain issues. ROC register is a catalogue that contains all magazines that have officially been included after a formal registration with the objective of guaranteeing a form of communicative transparency from the editors who should grant a correspondence of their published data by treating and ensuring the updates. In the catalogue it is possible to select from a range of source filters, such as, the juridical nature, the geographic organization's base, which type of activity the operators offer, and the numbers of their registration that confers a solid level of authoritativeness. In this preliminary filtering operation to select the main matches between the magazines and the topic technicalities proper to the Cybersecurity field of knowledge, apart from the semantic adherence a magazine can cover, other parameters have been taken into account, such as:

– **the authoritativeness of publishers**: just being registered on the ROC list can represent itself a seal of guarantee, but, sometimes, a parallel more accurate search on other portals to expand the query process can be useful to carry out;
– **the format**: digital or paper-based, in the case a paper-based magazines, conceived as specialized covering the domain's subject fields, the text recognition scanning tasks have been launched to digitize the textual data (e.g., *ZeroUno magazine*);
– **the period of time**: only magazines issues published from 2002 have been taken in consideration in order to be uniform with the legislative sub-corpus target;
– **the years coverage**: this feature refers to the fact that magazines should at least provide a minimum number of issues going back in time; hence, magazines with two-monthly publications, or at least monthly, per year have been preferred;
– **the language**: firstly, only magazines published in Italian language have been selected, or the bilingual ones that could allow an automatic equivalence matching procedure;

---

[1] http://ulrichsweb.serialssolutions.com/
[2] https://www.jstor.org/
[3] https://www.oclc.org/en/worldcat.html
[4] https://www.kit.edu/english/
[5] https://opac.sbn.it/opacsbn/opac/iccu/change.jsp?language=en
[6] https://opac.bncf.firenze.sbn.it/bncf-prod/
[7] https://www.agcom.it/elenco-pubblico

Ulrich's magazines		BNCF magazines	
magazines	search queries	magazines	
Fondamenti di Informatica	cybersecurity	1.-Hoepli informatica. Reti and sicurezza	
Informatica e Diritto		2.-Information security: testata specializzata sulla sicurezza informatica	
Sicurezza Digitale		3.Antivirus and sicurezza: il manuale della sicurezza informatica (2005-2011-2008)	
Il Diritto dell'Informazione e dell'Informatica		4.-Speciale Hackers magazine (2011)	
1981 Informatica e Ordinamento Giuridico		5.-Guida alla sicurezza informatica (2002-2003)	
Tecnologia e diritto 2017		6-Sicurezza informatica: la rivista della sicurezza globale dei sistemi informativi (1994-1999)	
		7.E-gov : metodi e strumenti di innovazione: pubblica amministrazione Italia-Europa (2001-2010)	
		8 Bancamatica : rivista mensile di elettronica, informatica e sicurezza nella banca e nella finanza(1996-2011)	
	cyber	1. Italian cyber security report: critical infrastructure and other sensitive sectors readiness (2013)	
		2.Information warfare (2010-2015)	
		3.Cyber: cervello, mente, coscienza: mensile di informazioni olistiche (1989-1997)	
	hacker	1. **Hacker journal (2002-2011)**	
		2.Play hacker : trucchi e codici per oltre 100 giochi!(2003)	
		3. Hacker attack magazine (2003)	
		4.Hacker 4ever (2002-2011)	

Figure 4.1: Magazines from Ulrich's web and BNCF for the Cybesecurity domain

– **the availability**: if the magazines resulted available in a free form online, they have been considered as candidate, but also the ones requiring a registration and a subscription if the topic focus was worth analyzing them.

The following Figure 4.1 provides a list of magazines obtained by searching through the catalogue of Urlich's Web classification and BNCF looking for cyber reference keywords, such as *cybersecurity, cyber* and *hacker*. It has to be noted that from the matches obtained from the Glossary of Intelligence entry terms and the titles of the magazines in ROC, few of them could have been considered adequate to fulfill the parameters for the magazines selection to be part of the sci-entific subcorpus. This is firstly due to the fact that the publication of the issues of many of them resulted to be interrupted, other times the magazines were not available in a digital form and the paper-based contents were accessible just for the current search year not covering enough information for the retrieval of the needed documentation; secondly, the topics resulted to be too much general, such as the magazine titles obtained by applying the keywords as *information technol-ogy, computer*. In detail, given the aforementioned specifications for the scien-tific corpus' compilation, the magazines resulted from the researches launched in the web portals by matching the terms of the Glossary of Intelligence with the

Table 4.1: Corpus criteria selection

Corpus criteria selection			
Coverage	**Language**	**Sources**	**Context**
2002–2017	Italian (*first phase*) and English (*alignment phase*)	Eur-Lex, Altalex, CERT, ENISA, ROC, ULRICH's, BNCF, experts repositories	National, European regional laws; multilingual Cybersecurity magazines

magazines titles, and among over 80 results, the ones selected for the purposes of this work are the following:
ZeroUno three-monthly magazine; *GNOSIS, Italian Intelligence Magazine* annual publications with 4 issues per year.

As Table 4.1 shows in bold, the other three magazines selected to be part of the scientific corpus are:
- *Informatica e Diritto (Information technology and Law)* with annual publications in 1 or 2 issues per time;
- *Diritto dell'Informazione e dell'Informatica (Information and Informatics Law)* with annual publications in 4 or 5 issues per time;
- *Hacker Journal*, a monthly magazine on the specific areas related to cyber crime, threats, hacking.

More specifically, the final source corpus, created after the selection of the domain-oriented sub-corpora following the discriminatory preliminary criteria for corpus creation, is constituted of **560** documents, as it can be observed by the following Table 4.2, and is made up of legislative documents and sector-oriented magazines, among these latter files the technical documentation given by reports and guidelines is included. Each category is subdivided into the following sub-categories per language in order to define the population for the bilingual corpora:

1. **Italian**:

   ■ Legislative dataset has been constructed by consulting the main open source regulatory portals, i.e., Eur-Lex, AltaLex, DeJure, and starting the queries by giving as input keywords belonging to the specialized field of knowledge, following the principles of selection based on time-range coverage, effectiveness, Italian language-oriented documents, semantic compliance with the domain's topic framework. It is partitioned in:

- **Regulations**, such as *"Regolamento (UE) N. 910/2014 del Parlamento Europeo e del Consiglio del 23 luglio 2014 in materia di identificazione elettronica e servizi fiduciari per le transazioni elettroniche nel mercato interno e che abroga la direttiva 1999/93/CE"* (Regulation (eu) no 910/2014 of the european parliament and of the council of 23 July 2014 on electronic identification and trust services for electronic transactions in the internal market and repealing Directive 1999/93/EC);
- **Norms**, such as "Norme in materia di misure per il contrasto ai fenomeni di criminalita' informatica. Legge 15 febbraio 2012, n. 12. (12G0027) (GU Serie Generale n.45 del 23-02-2012) "(*Law   15 February 2012, n. 12 Norms regarding the preventive measures to fight against cybercriminality phenomena (12G0027) (GU General Series n.45 del 23-02-2012))*
- **Laws**, such as *"Legge 18 marzo 2008, n. 48 Ratifica ed esecuzione della Convenzione del Consiglio d'Europa sulla criminalita' informatica, fatta a Budapest il 23 novembre 2001, e norme di adeguamento dell'ordinamento interno"* (Law 18 March 2008, n.48 Ratifying and executing the Convention of the Council of Europe on the cybercriminality, at Budapest 13 November 2001, and relevant rules of adaptation to the internal juridical system) – **Decrees**, such as *"Decreto del Presidente del Consigio dei Ministri 24 gennaio 2013 Direttiva recante indirizzi per la protezione cibernetica e la sicurezza informatica nazionale. (13A02504) (GU n.66 del 19-3-2013)" (Decree of the President of the Council of Ministers of 24 January 2013. Directive introducing guidance to cyber protection and national cybersecurity 13A02504) (GU n.66 of 19-3-2013)*

■ Scientific dataset is represented by the sector-oriented magazines issues focused on the Cybersecurity domain information. The five main domain-specific selected magazines in Italian have been chosen by using items keywords concerning the domain under study and the semantic correspondence that exists in each journal with respect to the knowledge field of interest [9]. The total number of the magazine issues is 314 and they have been picked up throughout the selected time range taken into account to construct the source corpus. They are all specific-oriented to the domain of Cybersecurity and security systems. The terminology inside is highly technical and targeted to the represent the specificity of the field of study. Together with the legal and the guidance documentation, the scientific journals sub-corpus supplemented the heterogeneity level of language of which the corpus should be characterized to better cover the variability of language to include in the semantic resources. The paradigm

to be observed in the compilation of the Italian corpus is to retrieve domain-specific language data that can be representative of the domain to be analalyzed, the datasets have to be published in Italian language and in Italy as geographic area. Therefore, the selection process of the issues meant to shape the scientific framework from which to extract technical terminology with NLP techniques have been preferred, as previously described, journals dealing with the specific topic of Cybersecurity in Italian language included within the official public register ROC[8] or in the list of the main magazine web-portals. To sum up, the following are all the magazines titles selected for the purposes of this research work:
- Hacker Journal[9] (2002–2004, 2006, 2008–2010, 2013)
- ZeroUno[10] (2015–2017)
- Informatica e Diritto (*Information technology and Law*) (2013–2017)
- Gnosis – Italian Intelligence Magazine (2013-2016)[11]
- CybersecurityTrends[12] (2017)
- Diritto dell'Informazione e dell'Informatica (*Information and Informatics Law*)[13] (2013–2016)

■ Technical documents:
- Guidelines, e.g., *CoE Electronic Evidence Guide. Version 2.1. (2014) CERT Guidelines* on Ransmoware, Social Networks, Mobile Devices, Cloud;
- Technical Reports, e.g., *Cyber Intelligence and information Security – CIS Sapienza Italian Cybersecurity Report (2016), OAD Osservatorio Attacchi applicativi in Italia Report (2017)* (*Observatory Applicative attacks in Italy Report, Clusit Reports on ICT Security in Italy (2013–2016, 2018–2019)*

2. English

■ Legislative dataset is the parallel corpus referred to the Italian one since the main web-portals which have been consulted for the compilation of the Italian legal documents framework provided the same in the English version for the normative context.

---

[8]https://www.agcom.it/registro-degli-operatori-di-comunicazione
[9]https://hackerjournal.it/
[10]https://www.zerounoweb.it/
[11]http://gnosis.aisi.gov.it/
[12]https://www.cybertrends.it/
[13]https://www.iusexplorer.it/Rivista/Diritto_dell_Informazione_e_dell_Informatica?area=riv_societa

■ Scientific dataset: to select the main journals to take into account to populate the scientific and technical terminology corpus, the consultation of domain experts resources represented a starting point from which to start selecting the most suitable issues for the domain. Examples of the selected English journals are:
- *Cyber Defense – CDM Magazine* (2013-2017)
- *Cybercrime and Espionage – An Analysis of Subversive Multi-Vector Threats* (2011)
- *Cybersecurity Symposium* (2016)
- *Cybersecurity Law* (2017)
- *Cybersecurity Lexicon* (2016)
- *Cyber Warfare – Techniques, Tactics and Tools for Security Practitioners* (2014)
- *The Cyber Risk Handbook: Creating and Measuring Effective Cybersecurity Capabilities* (2017)
- *The Basics of Cyber Safety – Computer and Mobile Device Safety Made Easy* (2015) – *Worldwide Cybersecurity Summit (WCS)* (2011–2012)

■ Technical documentation: also for what regards the digital documents constituting the technical collection, the English version is provided for the selected Italian ones.

Table 4.2: Italian Cybersecurity corpus size

	Number of documents	Number of terms
Laws	246	6 345 447
Sector-oriented magazines	314*	3 554 192

Among the 314* documents, which is the total number of issues each of the selected magazines has published, magazines contain these following partitioned data:
– **GNOSIS** has 4 issues *per* each year from the period taken into account 2013 to 2016;
– **Hacker Journal** has for 2002 – 18 issues, for 2003 – 23 issues, for 2004 – 4 issues, for 2006 – 2 issues, for 2013 – 1 issue, for 2008 – 26 issues, for 2009 – 21 issues, for 2010 – 11 issues;
– **Information and Informatics Law**, for the period going from 2013 to 2017 presents 5 issues per year;
– **Information technology and Law**, for the period going from 2013 to 2017 presents 1 or 2 issues per year;

– **ZeroUno**, presents for 2015 – 11 issues, 2016 – 9 issues and 2017 – 5 issues, plus 2 inserts;

– **Cybersecurity Trends**, for the year in which the corpus compilation has started, 2017, has one yearly publication constituted of 4 issues.

To calculate the statistical representativeness of the Italian Cybersecurity compiled corpus, Table 4.3 illustrates the number of tokens (all the words present in the source corpus), the number of types, that is the unique words (1-gram; 2-gram), and the number of hapaxes. Then there is a score for the Type Token Ratio (TTR) formula and one for the proportion of hapaxes in the vocabulary, meaning the words appearing just once in the source corpus, a measurement that provides a computation of specificity and accuracy of the selected documents. Here, concerning the corpus built for the Cybersecurity almost 70% of words occur once, the most frequent are, for instance: *risks, control, technologies, Internet, devices*, and in the first hundreds there is also included the term *Cybersecurity*. For what concerns the unique words retrieved with the 1-gram and 2-gram computation, the most frequent are – except for the stopwords which in this phase are provided in output and then will be removed by using more sophisticated term extraction techniques, respectively:

■ 1-gram: *security, services, protection, control* and *software*

■ 2-gram: *European Parliament, legislative decree, personal data, access to* and *electronic communications*

Table 4.3: Types Tokens Hapaxes in the source corpus

N. of tokens	N. of types	N. of hapaxes	TTR	Hapax
9 899 639	1-gram → 446 178 2-gram → 2 506 310	303 356	4,8%	67,98%

4.1.2 Software-aided terminological extractions – First selection

The candidate terms have been selected from the controlled lists the pre-trained software semi-automatically provided. The first selection of software took into consideration several multilingual tools, able to work firstly with Italian and then with English. Among these, the ones that have been tested on the source authoritative corpus are:

■ **T2K**[14]: is a native Italian linguistic tool meant to process big datasets both in Italian and in English, it has been the first software employed to retrieve

---

[14]http://www.italianlp.it/demo/t2k-text-to-knowledge/

the first controlled list of candidate terms. Among the others it proved to work efficiently over massive amounts of textual data, especially in Italian language, which is the native language the tool has been developed on. The following section 4.1.2.1 will describe its main features and the results obtained from the Cybersecurity corpus.

■   **Wordsmith**[15]: is a linguistic toolkit, specifically a Lexical Analysis Software launched by the Oxford University Press since 1996 and works for the detection of word patterns. As its main advantages there is the possibility to run operations with several languages. The drawback is that it is not open source. Nonetheless, it offers many operational tasks to execute on the textual corpora, such as, wordlist and keywords extraction, concordance, tagging. It applies the rules of minimum frequency and length of collocates to provide the terms lists, and the reference to the contextual information related to the documents given in input.

■   **Nooj**[16]: is a open source software that works following the principles contained in Chomsky's generative grammar, in particular the regular grammars. Its main functionality is referred to the annotation systematization of texts that can be customized by the users working with intra-word linguistic units. The texts given in input are exported in XML form. The recognition of different parts of texts leads to the generation of semantic recursive chains to be retrieved in corpora and it can become useful in the perspective of creating pattern-oriented rules. Also in this case, the drawback is the slow performance when it comes to treating very large datasets, and the time-consuming human-effort in creating semantic concatenations meant to become patterns to be found in the texts.

■   **LancsBox**[17]: developed at Lancaster University, is a software toolkit working with the linguistic analysis of textual corpora, it extracts wordlists from documents, creates ngrams, PoS tagging with TreeTagger, collocations which can be visualized in a graph, full context of search terms. The software supports multiple simultaneous analysis of linguistics tasks, saving concordances. A useful function is the split screen panel for two analyses to be run at the same time, and the possibility to overlook contemporary to categorized information. Again, the disadvantage is that this toolkit works rather slowly with many and long textual files given in input.

■   **AntConc**[18]: is a open source software to extract terminology from corpora and executes concordance tasks and text analysis. This software also

---

[15]https://www.lexically.net/wordsmith/
[16]http://www.nooj-association.org/
[17]http://corpora.lancs.ac.uk/lancsbox/
[18]https://www.laurenceanthony.net/software/antconc/

gives the possibility, as the aforementioned other ones, to retrieve the n-grams, wordlists and keywords list, but takes behind the same limit of slow performance in working with large datasets.

■ **Termsuite**[19]: is a linguistic toolbox working with six languages, and its main functionalities are related to terminology extraction, variants recognition, morpho-syntactic analysis and specificity computation. Its main advantage, that has lead to its selection over the other software, is its rapidity in providing outputs while processing conspicuous textual documents. In the following section 4.1.2.2, its main tasks will be presented with detailed descriptions.

From this list of computer-aided terminological extraction, just T2K and TermSuite have been selected because of their fast and precise performances and accuracy in executing very sophisticated linguistic processing operations over large corpora in Italian language.

4.1.2.1 T2K – Linguistic-oriented tool

T2K [61] is native Italian extractor developed by the Italian Computational Linguistics Group at Computational Linguistics Institution (ILC-CNR) in Pisa. It is based on the assumption that the most relevant concepts proper to particular reference texts are piloted by the most recursive terms. The software executes semantic extractions starting from domain-specific textual documents given in input to the system, and runs terminological processing operations by applying default or customized configurations giving in output a glossary or an index with semantic information on terms. In this way a collection of terms, that are structured according to conceptual cluster groups, allows a conceptual-semantic network development. The way by which T2K manages and structures the information extracted is through taxonomical concatenations, PoS tagging, documents indexing and dependency parsing functionalities.

Once completed the main basic linguistic pre-proccessing operations, T2k returns three types of controlled terms lists:
- terms sorted by applying statistical calculations, i.e., the inverse frequency measure: terms with the highest frequency are placed in the first positions, together with their numerical scores;
- terms indexing given by collecting the list of terms accompained with information about their positions in the sentences and in the documents;
- BT-NT list, meaning Broader Term-Narrower Term list, which disposes the most frequent terms as head-nouns grouped with co-occurring terms in the documents given in input, for example:

---

[19]http://termsuite.github.io/

hacker

-> hacker etico (*ethical hacker*)

-> hacker europei (*european hacker*)

-> hacker italiani (*italian hacker*)

-> hacker responsabili di violazioni di sistemi (hacker responsible of systems' intrusions)

virus

-> virus coder

-> virus infetta (virus infects)

-> virus informatici (informative virus)

-> virus macro

-> virus trojan

As it can be observed from these two head-based term lists, in the perspective of augmenting the procedure of semantic relationship detection automatization, using this configuration can help in identify the narrower terms of the most generic ones, that are conceived as heads of these groups, and, thus, facilitating the recognition of hierarchical structures. For instance, following this principle, the term *virus* will be the Broader Term BT of the compound *virus trojan*, which becomes respectively its more specific term NT. Once identified a hierarchical structure, terms have to be normalized for their inclusion in the thesaurus (Iso 25964-1:2011) in order to guarantee a level of compatibility with a unified shared knowledge. Indeed, *virus Ttrojan* has been validated by the experts of the domain in its form *trojan horse*, because of the multiple terminological usages of the second sense in official documents shared in their community. Among the main benefits T2K provides there is the leading developers group's nationality: the fact that it is an Italian spoken expert group makes the functionalities behind the system highly accurate for the corpora written in this language. And this falls within the purposes of the use case of this project activity, which supported the final user needs of creating a semantic resource able to orientate the understanding of a highly technical terminology referring to the Italian Cybersecurity framework, which will then be provided in a English version too. The development of terminological extraction techniques and the automatization of the semantic means of control are meant to be adapted as methodology to other fields of knowledge.

The terminological accuracy is provided by a task included in the software toolbox, that is the application of a contrast corpus, which can be pre-loaded by the system as the generic language to be compared with the most specialized corpora under processing executions, or presented as the lists obtained by previous terminology extractions to be mapped with. This allows a comparison between the controlled list of terms extracted from technical corpora and a more generic wordlist with a computation score of precision, making the two terminological sets comparable for statistical analysis. The contrast guarantees the identification of the most representative terms of the domain under analysis since it modifies the positions of each term included in the list with respect to its frequencies comparison in the two lists, highlighting in this way the higher level of specificity or generality. T2K offers the possibility to manually operate modifications over the pre-imported morpho-syntactic rules to be applied on the source corpus to extract terms. This means that users can manipulate the syntactic and statistical configurations provided by the software to achieve the desired outputs. For instance, the software allows the customization of patterns that are to be used for the extraction of domain-oriented multi word-unit terms, showing a high level of flexibility in the way users can apply the POS-tagged configurations settled down to the terminology needs that can be stored in the back-end system for other future input corpora exhibiting the same properties. For what concerns the Cybersecurity domain, the terms have been extracted by applying a morpho-syntactic configuration that has given as output single and multi-word lexical units; the maximum terms length has been fixed from 1 to 6 grams, e.g., *hacker* (1 gram), *sistema di crittografia a chiave pubblico (public keys criptography system)* (6 grams), the frequency has been put equals to minimum 1 in order to retrieve all the occurrences in the corpus. The semantic configuration established to run the terminology extraction starts with a NOUN, followed alternatively by a complex or simple PREP, an ADJ or a NOUN, and ends with a NOUN or an ADJ. In detail, applying the range of modifications T2K allows in the patterns, the total numbers of entries (single terms and MWTs) is 593,888 – with the highest value of frequency equals to 35271 (*data*) –, a score that reveals the high level of possible noise that can be found among the records. The elevated number of terms given in output is probably due to the fact that T2K does not allow the possibility to insert an external stopword list, or a set of variation rules like TermSuite, as it will be noted in the next section, that could help in filtering out undesired lexical items and reduce the noise in the outputs. Indeed, in the results provided by the software there is a number of invalid expressions, partially correct grammatical phrases that do not represent lexical units carrying with them meaningful values to be integrated in the term lists. Some of the candidate terms have been, for instance, excluded from the selection since they appear not to be meaningful bearers with respect to the domain of Cybersecurity (e.g., *article, information, member, level and work*) or proper to other fields of knowledge akin to the one under analysis (e.g., *legislative decree, law, comma, ownership and treatment*). The configuration of low frequency and high levels of terms combination has aimed at achieving a comprehensiveness in the terms selection. To reach this goal the T2K term lists have been obtained with the application of this configuration structure:

1. the minimum threshold frequency has been set to level 1;

2. the morpho-syntactic chains have been defined as Common Nouns (S) followed by an Adjective (A), or/with a a Preposition (E), Articulated

Preposition (EA), Common Noun (S), and ending with a Common Noun (S) or an Adjective (A); POSStart p:S

POSInternal p:EA p:E p:A p:S

POSEnd p:A p:S

In particular, the resulting terms reflect these main chains:

- Noun (S) *hacker*
- Noun + Noun (*cybersecurity, piattaforma internet – internet platform*)
- Noun + Adj (A) (*firma digitale – digital signature*)
- Noun + Adj + Adj (*crimine informatico economico – economic cyber crime*
- Noun + Adj + Noun (*virus informatico vero – true informative virus*)
- Noun + Prep (EA) + Noun + Adj (*accesso ai dati personali – access to personal data*
- Noun + Adj + Prep + Noun (*attacco informatico al sistema – cyber attack to system*)
- Noun + Adj + Prep + Noun + Adj (*accesso abusivo a sistema informatico – illecit access to informative system*

It has to be underlined that the multi-lingual intrinsic nature of the Cybersecurity domain leads to the visualization of a range of terms in English forms, representing borrowings, i.e., "the result of the transfer of a linguistic phenomenon from one language for use in another" (Daille 2017:21). Therefore, both in T2K and TermSuite output lists there are many occurrences in the original English version, as for example *hacker, virus, honeypot, bitcoin*, etc., that have been integrally implemented in the target language acquisitions.

Nonetheless, despite the fact that T2K offers multiple functionalities in working with unstructured textual data models to execute operations of text mining, linguistic analysis, semantic richness calculation by applying a contrast corpus, it does not allow a truly expanded set of pattern-based interventions in the rules and it shows protracted working time-interval to provide the final output.

4.1.2.2 *TermSuite*

Given the reasons described above, another software has been employed to enrich the semantic extraction towards the selection of the best accurate and domain-specific candidate terms to be included in the thesaurus' configuration. TermSuite [46] is a term extractor tool written in JAVA language, multilingually designed and extendable to the detection of variants starting from a source corpus. The software works in computing the termhood and the unithood of a term candidate (Kageura, 1996). The modalities by which TermSuite toolkit processes the corpus basically rely on the pre-processing tasks of tokenization, PoS tagging, lemmatization, stemming, and morphological compounds detection. The post-processed dataset then undergoes a recognition of MWTs units that will be grouped in variants extracted by using Token Regex, which allows to establish a set of rules on the sequence of Unstructured Information Management Architecture (UIMA) framework annotation, every rule is conceived as a regular expression. As Daille (2017:209) points out:

> 'A rule formulated in UIMA Token Regex rules language consist of:
> 1. key word `term`
> 2. rule name (string expression between double quotes ("))
> 3. separator :
> 4: annotation utterance written as a regular expression,
> 5. semicolon ;
> [...] Rule an below extracts MWT composed of one or several adjectives followed by a noun.
> `term "an" : [category=="adjective"]+ [category=="noun"];`
> Rule an, below, extracts terms of A N pattern, such as *blade radius*.'

The terms can be systematized according to terminology operation of specificity computation and C-Value scoring. TermSuite executes the terminology extraction process by identifying and collecting term-like units in the texts, a subset of nominal phrases, filtering them and then sorting candidate terms with respect to the unithood, the terminology specificity and the level of adherence to the target application final needs.

The main benefit deriving from using TermSuite is the recognition of variant complex units in the text that has represented a methodology to retrieve in a semi-automatic way a set of semantic relations to be inserted in the thesaurus for the Cybersecurity domain. Indeed, the variants detection process resulted highly efficient to enhance the first structure of the thesaurus since the application of customized JAVA rules acts on the improvement of the accuracy in the term lists.

TermSuite customization for Italian

As mentioned above, the extraction of candidate terms, as well as their variants, is executed with the Token Regex expressive language, which enables users to delineate morphological and syntactic rules expressed over the annotations taken from `Stanford TokensRegex` (Chang and Manning 2014) [25]. The variation patterns are detected thanks to the system developed inside TermSuite, UIMA Tokens Regex, according to which the multiword units are ranked with "the most popular termhood measure" (Cram, Daille 2016). Unlike T2K, TermSuite gives the possibility to set customized patterns that enable the extraction of desired lexical outputs.

In the extraction provided by the TermSuite toolbox, in order to visualize and analyze the variations' configurations, both on the syntactic and morphological level, the process of alignment with the Italian grammar structuring rules has been necessary to be applied on the source corpus. In particular, the Italian language rules systematization in TermSuite has been adjusted in the morphology package (i) beginning with the inclusion of fixed expression lists, (ii) adding a general language support vocabulary, (iii) synonyms banks, (iv) executing modifications of prefixes, suffixes and suffixes derivation banks, (v) and root banks (vi), (vii) then applying new multi-word token regex rules for Italian, as well as (viii) the variants configuration to be adapted to this language-oriented Cybersecurity corpus. For what concerns the prefixes lists of Italian constructs, some of the main ones that have been added are:

agro- an- a- ad- ana- ante- anti- anto- arci- auto- avan- bis- **cyber-** de- di- dis- di- epi- eu- extra- fra- re- with respective exceptions, which are properly words, references such as:

- prefixes 'de-' -> e.g., decomposizione (*decomposition*); 'tra-' -> trasformazione (*transformation*); 're-' -> reazione (*reaction*).

For what regards the suffixes, modifications have been integrated in the derivation banks that allowed to have several variants in the terminology extraction, i.e., some nouns can be under different forms and from an adjective a noun can be derived and viceversa:
>ore
N N -ore -o -ore -a
>ere;-iere
N N -ere -a N N -iere -a
>ero
N N -ero -o N N -ero -ia
>anità
A N -anità -ano
>arezza
A N -arezza -aro
>ità
A N -ità -o A N -ità -e
>ezza
A N -ezza -o
>eria
A N -eria -ere A N -eria -o

In this terminology software there are five grammar variants recognition configurations for five languages, differently from T2K that works just for English and Italian. These grammar systematizations collect rules referred to the morphosynstactic structures according to the selected languages and the main three typologies of variants the software detects over large corpora, i.e., denominative, conceptual and linguistic [31]. As Daille (2017:37) affirms, a variant

> "is characterized by the following properties:
> 1. a variant always involves at least one term;
> 2. a variant is obtained by applying at least one linguistic operation which belongs to a mechanism for denominative and conceptual variants;

3. a term can produce several variants;

4. the number or utterances of the term in a text is slightly superior to the number of utterances of the variant. Equality of utterance numbers may be encountered for graphical and denominative variants."

Denominative variants, as the author highlights, are piloted by a range of construction mechanisms such as:

■ synonymic substitution of morphemes, functional words or lexical items with same language elements

■ symplification, modifying term patterns removing components, it improves the term conciseness. This mechanisms is, in turn, subdivided into other two techniques:

    1. **compression**: maintenance of the terms' lexical elements while moving one of them in another shorter one; e.g., hacker del telefono N P N 1 (*hacker of the phones*) → hacker telefonico N A (*phone hacker*)

    2. **reduction** : removal of one of the terms' lexical elements in a complex term referring to the same concepts, e.g. sicurezza del sistema informativo N P N A (*security of informative system*) → sicurezza di sistema N P N (*system security*). This can help in the decision-making process for the hierarchy relationship detection.

■ **exemplification**: modifying, decompacting term patterns by adding components, e.g., documento di strategia (*strategy document*) N P N → analisi del documento di strategia N P N P N (*analysis of the strategy document*).

Conceptual variants are the enhanced version of the denominative variants in the sense that the mechanism of expansion is favoured in describing specific terms both at the morphological and syntactic level. Among the main techniques covering the terms expansive conceptualization there are the ones referred to:

■ **derivation**: is a process of deriving a term from another, in particular by using prefixes and affixes, e.g., crittografia simmetrica N A (*symmetric cryptography*) → crittografia asimmetrica N A (*asymmetric cryptography*). The main output is the nominalization, for example, *bullies* → *bullism*. This can help in the decision-making process for the associative relationship detection;

■ **juxtaposition**: in initial or post-position when certain elements are put in initial positions or moved on the right (or left) of the term with modifiers, e.g. firma elettronica N A (*electronic signature*) → firma elettronica qualificata N A A (*qualified electronic signature*). This can help in the decision-making process for the hierarchy relationship detection;

■ **insertion** when the linear sorting outline of terms is edited by the insertion of other elements, e.g., attacco bruto N A (*brute attack*) → attacco a forza bruto N P N A (*brute force attack*) This can help in the decision-making process for the equivalence relationship detection.

Finally, the **linguistic variants** are related to the level of language, in particular the phenomena of:

■ **graphical and spelling**, supporting the synonyms identification, which can involve:
- the space structures, such as *cyber crime* → *cybercrime*;
- the hyphenation process, such as *cybersecurity* → *cyber-security*;
- the American and British spelling;
- writing agreements, as for example the capitalized letters in certain languages, as French, lose the accents of the terms;

■ **inflection** refers to the gender, plural, numbers for nominal simple terms.

The process of re-engineering the patterns configuration for the Italian language has implied a restructuring of the semantic chains this language properly shows, the base-terms [28]. These concatenations in Italian are syntactic structures mainly constituted by
Noun Adjective, e.g., sicurezza informatica *(cybersecurity)* **Noun(Prep(Det)) Noun2** sicurezza della rete *(network security)*.

Each of these basic patterns of base-terms can be modified by using the head part of these latter and clustering the occurrences found in the source texts.

Variants detected by using Termsuite software have represented a key point in the retrieval process of the hyponyms referred to the main candidate terms of the thesaurus and their synonym substitutes to be included. Additional examples of the variant structures found with the application of customized patterns in TermSuite are the following:

1. **Morphological variants**
   1) T n: homepage
   V[m] nn: home page

   2) T n: antivirus
   V[g] a: anti-virus

   3) T npn: hacker del telefono (phone hacker)
   V[d]+ na: hacker telefonico

2. **Syntactic variants**

1) T npn: sicurezza della rete (network security)
V[s] npncpn: sicurezza della rete e del sistema (network and system security)

2) T na: indirizzo elettronico (electronic address)
V[s] npna: indirizzo di posta elettronico (mail address)

3) T na: fascicolo sanitario (personal health record)
V[s] naa: fascicolo sanitario elettronico (electronic personal health record)

■ **hacker (519 matches)** → hackeraggio (*hacking*)

■ **worm (102 matches)** → worm via posta elettronico (*worm via e-mail*)

■ **antivirus (127 matches)** → antivirus affetto da trojan (*antivirus affected by trojan*)

4.1.2.3   Pke-keyphrases detection

Pke (Boudin, 2016) is an open-source python keyphrase extraction toolkit which implements a range of keyphrase extraction approaches, i.e., TopicRank, MultipartiteRank, TF*IDF and TopicalPageRank. Pke's architecture starts with reading a set of textual documents, then selects the candidates to be considered as keyphrases, in this second step two approaches are executed: the first one is a unsupervised approach that begins with the candidate weighting measure and then operates the classification; the second is supervised and starts by extracting the features toward the subsequent candidate classification; the N-best solution represents the last passage for the keyphrases output (Boudin, 2016). This library is multilingual and mainly relies on statistical features on a document oriented extraction level. The main scope by which this toolbox has been employed is the extraction of information regarding the topic terms, single or multi-word, which have connoted the textual data in the source corpus, with the objective to use them in order to check the thesaurus structure of semantic relationships.

The outputs provided by MultipartiteRank and TopicRank over the 560 documents inside the Italian Cybersecurity corpus, which consistently achieved the best scores for the terminology extraction task, gave 15 keyphrases per each document – going further would yield more noise from a keyphrase extraction point of view and for manual investigation efforts –, with a total number of scores equals to 8269. The results of terminology extraction for Pke methods according to the reference « train » corpus (training only occurs for TF*IDF and TPR (TopicalPageRank) which uses LDA), and top number of keyphrases from each document used for evaluation (–1 for all 10000 extracted keyphrases) show that when using all keyphrases each approach exhibits the same results. The maximum Recall achieved is around 40% when using the top 1,000 keyphrases, the Precision overall is less than 1%. Therefore, Pke proves not to be highly efficient for terminology extraction, but from a linguistic point of view it represents a reliable and accurate technique to be employed for the semi-automatic structuring of information because multiple candidate terms, having undergone a selection process to be included in the Italian Cybersecurity thesaurus, are gathered together with others, and this can support the process of associative semantic network identification as well as help through the automatic detection of the main top categories to be included in the thesaurus. Indeed, Pke library supported the definition of topical information about the domain and the keyphrases obtained with the application of its models guided the selection of the semantic clusters to be structured in the form of hierarchical systematization in the thesaurus, and consequently in the taxonomic framework of the ontology. The obtained clusters provided additional, or at least target-oriented, kind of information to structure the specialized terminology of the Cybersecurity domain. The following examples offer an overview of the main semantic clusters which have been retrieved by using Pke library with a partitioning when they belong to legislative, popular or technical datasets:

■ **cybersecurity**

- ■ in legislative documents: sistemi (*systems*), sicurezza (*security*), attacchi (*attacks*), rete (*network*), analisi (*analysis*), rischi(*risks*);

- ■ in popular documents: is found with the following keyphrase groups: sistemi *systems*, dati (*data*), cybersecurity, sicurezza (*security*), dispositivi (*devices*), protezione (*protection*), rischio (*risk*), software, attacchi (*attacks*), cyber-insurance;

- ■ in technical documents: sicurezza informatica (*cybersecurity*), app, smartphone, dispositivi (*devices*), email, dati (*data*), accesso (*access*), social network.

■ **hacker**

- ■ in legislative documents: not occurring, proximal compounds are cyber crime, cybercriminality;

- ■ in popular documents: sito (*website*), sistemi (*systems*), rete (*network*), sicurezza (*security*), attacco (*attack*), worm, software, password;

- ■ in technical documents: cyber space, attacco (*attack*), rete (*network*), computer, sicurezza (*security*), virus.

■ **cyber attacks**

- ■ in legislative documents: misure (*provisions*), sicurezza (*security*), controllo (*control*), pubbliche amministrazioni (*public administration*) and sicurezza ict (*ict security*);

- ■ in popular documents: cyber attack is found with sicurezza (*security*), spionaggio (*espionage*), vulnerabilità (*vulnerabilities*), cyber-security, protezione (*protection*), difesa cyber (*cyber defence*), information agency, rete (*network*), nuova password (*new password*);

- ■ in technical documents: sicurezza (*security*), sistema swift10 (*swift10 system*), sistemi(*systems*), analisi (*analysis*), rischi (*risks*).

■ **virus**

- ■ in legislative documents: sistemi (*systems*), controllo (*control*), gestione (*management*), sicurezza (*security*), pubblica amministrazione (*public administration*), material handling, dati (*data*);

- ■ in popular documents: sistemi(*systems*), dati (*data*), rete (*network*), applicazioni (*apps*), iphone, diritti (*rights*), hacker, computer, password.

From these grouping configurations it appears that in most of the cases the legislative documents exhibit a more generic and politics-oriented terminology, e.g., *control*, *management* and *public administration*, while in the sector-specific magazines and reports or guidelines, which compose the popular-scientific dataset, the terms are more precise and accurate with respect to the domain to be analyzed.

This is a useful conclusion in the observation of the corpus' composition, in the way its heterogeneity in the documents selection can guarantee a more expanded cross coverage of the domain.

Another benefit obtained in executing Pke models, is linked to new semantic structures given in the clustering techniques application. Indeed, the list of clusters coming from the PKE models implementation sometimes presented new terms that did not appear in the previous terminology extractions, and together with their cluster sets have been a valid means by which the semi-automatic structuring of the semantic connections network inside the thesaurus' outline has been enhanced. For instance, the candidate complex term *cyber counterintelligence* cluster, has been highlighted as new term to be considered as candidate and its proximal semantic components with which it occurs provide new semantic connections that can become marked up items for classified information in both thesaurus and ontology: *technique* and *procedure* in the cyber intelligence and cyber defence tasks can be conceived as labels to create new structured semantic chains.

attività (*activities*) intelligence, controspionaggio (*counter espionage*), tecniche (*techniques*), **cyber counterintelligence**, cyber actions and difesa (*defence*)

Every cluster presents its own weight size per each model, for example the outputs given by executing TopicRank algorithm associate the main topics retrieved in the corpus with their weight scores, and this is a trigger towards the accuracy information related to terms:

1. ('sistema informatico,' 0.03343072271898668) (*informative system*)

2. ('reato,' 0.01763567632052654) (*crime*)

3. ('dati,' 0.014879050118011557) (*data*)

4. ('accesso illegittimo,' 0.014062895607557649)(*illegitimate access*)

5. ('accesso abusivo,' 0.00897217015528124) (*abusive access*)

6. ('rete,' 0.005324747843856192) (*network*)

7. ('sito,' 0.0038876214378223196) (*website*)

8. ('server,' 0.0036813516347631203) (*server*)

4.1.2.4 Results

TermSuite and T2K, the two term extractor software, provided different candidate terms extractions both from a quantitative and qualitative perspective. More specifically, the TemSuite whole term lists results obtained over the same corpus as T2K, resulted to be lower than the ones from T2K, this can be due to the fact that the implementation of the Italian morphological package, the configuration of prefix banks and derivational datasets, variants rules detection through pattern-based approaches, and supporting external vocabularies of synonyms in Java for Italian in the back-end system, as well as the customization of stopword lists, allowed to execute more sophisticated algorithms in retrieving precise terminology avoiding semantic noise (like non-sense terms, repeated lexical units, etc.)

T2K presents more terms in the list extracted from the source corpus, and this can be either because of the high maximum terms output scores and for the length in the terms grams, or for the lack of more specific internal adjustments as TermSuite. Unlike TermSuite, which returns a very cleaned up list of terms in the vocabulary, the overall number of terms in T2K sometimes presents some inconsistency with the semantic preciseness and adjustments are needed by terminologists after the domain experts' validation of the candidate terms. Probably,

Table 4.4: Term extraction details

	T2K	**TermSuite**
Number of candidates	593 887	17 083

the lack of accuracy in the correctness of the terms is attributable to the impossibility of inserting a customized stopwords list that could avoid having such inconsistent terms and applying personal syntactic and morphological rules, as TermSuite, that could enhance the terminology filtering in the scores.

Notwithstanding, T2K has as one of its main features, which has helped out in the semi-automatic detection of the semantic hierarchical relationships, the output in the form of glossary of terms gathered by head-based terms. The Italian software in its linguistic processing collects the resulted terms from the extraction in a grouping configuration putting under the ones occurring most of the times in the source texts, i.e., head-based terms, the co-occurring others in a chained-linked structure: the inference system created encapsulates a term with high frequency score with many others it occurs together within the corpus. By the means of this application approach provided by T2K, the terminology procedure will have as many MWTs as possible to re-create a hierarchical system; for instance, a term like "Hacker" before or after a single term considered as the head one.

This implementation is executed in TermSuite by creating clustering with respect to the variant type, thus, the list of terms coming from the terminology extractions provides a terms gathering by principle of variation detection.

To verify the level of terms granularity, namely the accuracy and domain's compliance of the representative candidate terms, the procedure that has been tested is a comparative structure with the Italian thesaurus for Cybersecurity, a tool that has been evaluated from the field of study experts. The process of validation applied to the thesaurus regarded the mapping association with the main gold standards of the Cybersecurity sector, among which NIST 7298 r2 vocabulary and ISO 27000:2016 taxonomy. For this reason, the terminological framework can be reasonably judged as a reliable structure with which juxtapose the results from the term candidates extraction software.

4.2   Candidate terms selection

4.2.1   Frequency criterion

The first measure that has been considered for the first selection of candidate terms meant to be included in the Italian Cybersecurity thesaurus and then in the ontology taxonomy, is the Term Frequency*Inverse Document Frequency formula, TF*IDF, that has also been exploited in the contrast cross-analysis with the more general language-oriented datasets. In particular, both T2K and TermSuite gave together with each term the frequency scores, the first software only provided the domain relevance score and the frequency of the terms with respect to the documental context, and provides, as function, that of applying a contrast corpus to detect the specificity of the terminology extraction. On the other hand, TermSuite offers a range of parameters, related to frequency computation, to be run over a large corpus as pipelines, such as:

- document frequency: number of textual data inside a source corpus where the term occurs

- frequency norm: term occurrences each every 1000 words inside a corpus

- general frequency norm: term occurrences each every 1000 words inside a general language corpus

- specificity: the term in the source corpus specificity compared with the general language one, i.e., weirdness ratio

- frequency: terms occurrences in the input corpus

- independent frequency: term occurrences in corpus exactly in their original form without variant attributed by the software processing system

- tf-idf: ration between frequency score and document frequency score

- spec-idf: ratio between specificity score and document frequency score

- semantic-similarity: the threshold specifiying the similarity for the semantic group in the monolingual contexts alignment

To calculate the specificity of given terms, meaning the termhood measure of the terms contained in the corpus with respect to the ones that appear in the Italian general corpus, statistical inner algorithms are executed. This statistical computer-assisted measurement is obtained by relating the *ratio* for two term lists (one specific for the domain, and another for the general documentation) obtained by calculating the *relative frequencies* for each of them. Figure 4.2 depicts the scores of several terms contained in the terminology extraction with a more generic list of Italian terms extracted from a general language corpus to underline how some of them are characterized by a higher level of specificity that allowed to select them with respect to others as entry terms in the thesaurus and ontology. Results proved that some of the key representative candidate terms, selected in accordance with the groups of experts' validation, exhibited a higher score in the specificity as for example *spam* or *malware*, others are retrieved very frequently in generic contexts and impacted on the termhood score, that is

term	frequency specific	frequency general	ratiofreqsp	ratiofreqg	specificity
cybersecurity	238	0.1	0.00011	1.81E+14	60030,79
virus	1185	5887	0.005	0.0011	5.077167
cyber threats	5	0.1	2.28E-06	1.81E-09	1261.15
cyberwar	3	0.1	1.37E-06	1.801E-09	756.69
privacy	2083	0.1	0.0009	1.81E-09	525395.5
spam	201	0.1	9,07E-005	1.81E-09	50698.27
spamming	45	0.1	2.05E-005	1.81E-09	11350.36
attacker	56	0.1	2.55E-05	1.81E-09	14124.9
phishing	136	0.1	6.2E-05	1.81E-09	34303.4
security	17365	28387	0.008	0.00052	15.42951928
access	8392	18728	0.004	0.0003	11.30241319
bitcoin	23	0.1	1.05E-05	1.81E-09	5801.295
malware	449	0.1	0.0002	1.81E-09	113251.4
protection	10327	15112	0.005	0.0003	17.23651008

Figure 4.2: Results of termhood measures

the case, for instance, of *virus*, *security*, *access* and *protection*, which are used very often in broad-sense discourses. It is for this reason that a corpus that is made up of highly specialized source documents is likely to provide more accuracy in the selection of the terms meant to be part of a semantic monitoring tool for technical domains of study, and this can provide a trigger for the evaluation of the terminological resources representativeness with respect to a given domain to be semantically managed and represented.

4.2.2   *Mapping with gold standards*

The procedure of selecting the most representative terms for the domain of Cybersecurity has been carried out also with reference to the main gold standards about this specialized field of knowledge, meant to be mapped with the corpus compiled to begin the selection process. Table 4.5 illustrates the numerical population of terms occurring in the Italian Cybersecurity corpus, *Cyber*, the ones present in the thesaurus created with the approval of the domain experts, *Thes*, the five reference frameworks with which the mapping process has been executed: the White Book of Cybersecurity published by the CINI Consortium in Italian language, *WhiteBook*, the glossary given inside the National Institute of Standard Technology (NIST), *NIST*, and the taxonomy in the standard ISO/IEC 27000:2016 (Information technology – Security techniques – Information security management systems – Overview and vocabulary), *ISO*, originally in English language and manually translated with the support of IATE (European Unionon of Terminology termbanks system[20], the Glossary of Intelligence published by the Italian Presidency of Ministers, *GLOSS*, and the summary of the glossaries

---

[20]https://iate.europa.eu/

contained in the Italian Clusit Reports from 2016 to 2019 *Clus*. Subsequently, table 4.6 provides an overall visualization on the common knowledge shared firstly between the terminology contained in the corpus, given by the term extraction processing software lists – the list provided by TermSuite has been preferred since it did not present noise in the outputs, such as orthographic errors or noisy symbols – with the lists of standardized terms in the gold standards, and secondly the terms filtered out in the first phase of selection with the Cybesercurity group of experts included in the Italian thesaurus with the same gold standards. In this way, the scores have identified the level of terminology coverage both the source corpus and the first beta version of the thesaurus have been able to grant with respect to the latter list of terms present in the thesaurus, which is characterized by the terminology units both included in the corpus from which the documents have been pre and post processed by the term extractor toolboxes and also by the additional information that the supervision and collaboration of domain experts provided.

Table 4.5: Reference quantitative population of main multilingual Cybersecurity gold standards

	Cyber	Thes	WhiteBook	NIST	ISO	GLOSS	Clus
n. terms	17 083	245	14 635	446	88	266	5407

Table 4.6: Terminology mapping results source corpus with the gold standards NIST, ISO, WhiteBook, and Glossary of Intelligence (GLOSS)

CyberThes	CyberNIST	CyberWhitebook	CyberISO	CyberGLOSS	CyberClus
55	28	3392	39	58	1701
"	ThesNIST	ThesWhitebook	ThesISO	ThesGLOSS	ThesClus
"	12	80	11	10	52

Much more in detail, the main mappings retrieved – along with their variants – between the source corpus term lists obtained through semi-automatic extractions and the gold standard are the following:

- **Cyber and Nist** : 'malware,' 'hacker,' 'backup,' 'audit,' 'protocol,' 'access,' 'gateway,' 'backdoor,' 'buffer overflow,' 'audit log,' 'attack,' 'cryptography,' 'proxy server,' 'access control list (ACL),' 'authentication,' 'cipher block chaining,' 'digital signature,' 'identity proofing.'

Terms resulting from this mapped cross reference are more related to technical information, or intrusion related event, as *backdoor* or *attack*.

- **Cyber and ISO**: 'sicurezza informatica'(*cybersecurity*), 'documento informatico/ documentazione informatica'(*digital documentation*), 'autenticazione' (*authentication*), 'crittografia' (*cryptography*), 'crittografici'(*cryptographic*), 'dato identificativo'(*user identification data*), 'spam,' 'antivirus,' 'audit,' 'chat,' 'valutazione del rischio'(*risk evaluation*), 'fatturazione elettronica'/'fattura elettronica'(*electronic invoice*), 'sistema informatico'(*informative system*), 'bluethooth,' 'vulnerabilità'(*vulnerabilities*), 'rischio residuo' (*residual risk*)

The type of semantic information the security management system vocabulary covers mostly concern the Cybersecurity methodologies, such as *cryptography, user identification data, antivirus, risk evaluation, vulnerabilities*.

- **Cyber and White Book**: 'supervisione strategica' (*strategic supervision*), 'pc infetto'(*infected pc*), 'tentativo di attacco'(*attack attempts*), 'criterio oggettivo'(*objective criterion*), 'crittografia asimmetrico'(*asymmetric cryptography*), 'strumento informatico' (*digital means*), 'social media,' 'sistema operativo'(*operative system*), 'account,' 'sicurezza'(*security*), 'infrastruttura digitale'(*digital infrastructure*), 'assessment,' 'risposta all'incidente'(*incident response*), 'social network,' 'sicurezza online'(*online security*), 'trattamento del dato personale'(*personal data treatment*), 'sicurezza del sistema informativo'(*nformative system security*), 'hacker etico' (*ethical hacker*), 'sicurezza cibernetica'(*cybersecurity*), 'identità digitale'(*digital identity*), 'hacking,' 'codice malevolo' (*malicious code*), 'injection,' 'sicurezza digitale'(*digital security*), 'cybercriminali' (*cyber criminals*), 'tecnica di attacco'(*attack technique*), 'privacy,' 'spazio cibernetico'(*cyber space*) 'spam,' 'exploit.'

The knowledge represented by the guideline book for Cybersecurity domain is referred to the main strategies pursued to overcome cyber threats and attacks events in the institutional contexts, it can be observed through terms as *objective criterion, strategic supervision, exploit, spam, malicious code*.

- **Cyber and Clusit**: 'tentativo di attacco'(*attack attempts*), 'protezione tecnologico' (*technology protection*), 'assessment,' 'risposta all'incidente' (*incident response*), 'attaccato' (*attacked*), 'sistema crittografici'(*cryptographic systems*), **'integrità del dato'** (*data integrity*), 'virus,' 'intercettazione'(*interception*), 'hacker etico' (*ethical hacker*), "cyber security,' **'dato personale'** (*personal data*), 'spionaggio'(*espionage*), 'denial,' 'sistema di protezione' (*protection system*), 'discrezionalità' (*discretion*), **'dato attendibile'** (*reliability*), 'sicurezza cibernetica'(*cyber security*), 'identità digitale'(*digital identity*), 'protection,' 'hacking,' 'codice malevolo'(*malicious code*), "accesso remoto'(*remote access*), 'sicurezza digitale' (*digital security*), **'protezione del dato personale'** (*personal data protection*), 'cloud,' 'attività illecita'(*illecit activities*), 'vulnerabilities,' 'privacy,' **'furto di dato'** (*data breach*), 'threat,' 'email,' 'sicurezza informatica' (*cybersecurity*).

Clusit reports concern the analysis of the most representative cyber crime events having a high level of critical aspects, so the included terms reflect this objective, such as, ethical hacker, *threat, attacks.*

-**Cyber and Glossary of Intelligence**: 'sicurezza informatica' (*cybersecurity*), 'infrastrutture critiche' (*critical infrastructures*), 'autenticazione' (*authentication*), 'crittografia' (*cryptography*),'cifratura'(*encryption*), 'sicurezza cibernetica' (*cybersecurity*), 'minaccia cibernetica' (*cyber threat*), 'analisi del rischio' (*risk analysis*), 'compromissione' (*undermining*), 'intelligence,' 'spoofing,' 'integrità' (*integrity*), 'cyber space,' 'minaccia' (*threat*), 'spionaggio' (*espionage*). Being the Glossary of Intelligence a means published by political organisms, the level of semantic specialization is institutionally-oriented, meaning that the terms sorted in the glossary bear a more governmental weight. They are mostly referred to undermined systems categorization through *espionage* techniques, threats, and protection methods: *risk analysis, cryptography authentication integrity*

Once having performed the cross mapping analysis with the gold standards, the most representative candidate terms have been taken and organized in order to build a more precise structure of the Italian Cybersecurity thesaurus in line with the official domain knowledge shared within experts, and consequently the classes in the ontology system.

The following Figure 4.3 collects the information provided by the mapping implementation algorithm that guided the selection of the entry candidate terms in four groups. The first category refers to the terms covering the cyber attacks information, the second is associated with **agents'** roles in cyber attacks and episodes, the third relates to the more **generic terms** about Cybersecurity, and the fourth addresses to the **protection techniques** carried out in the Cybersecurity field of knowledge.

Among the main gold standards, the Clusit term list proved to be more adequate for the cross-mapping systematization both for the same source language (ita) and for the technical terms chosen, therefore, an evaluation over the five terminological lists, *Clusit, Glossary, Nist, Iso* and *Cyber* and on two corpora: Clusit (*Clusit*) and Cybersecurity (*Cyber*) has been executed to find out the measures referred to the terminology coverage. The results, in Table 4.7, are provided considering the Precision (P), Recall (R) and F-measure (F1), and the coverage of each list on every corpus is given too. Another method of terminology extraction that has been tested for the purposes of this classification is the Bidirectional Encoder Representations from Transformers (BERT), a recent deep neural network methodology, remarkably performing among several downstream NLP tasks [84], such as, next sentence prediction, question answering, name entity recognition (NER), and it also can be exploited to execute feature extraction or for classification, as a binary classifier for term prediction.

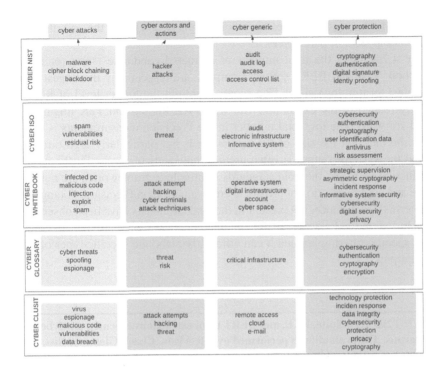

	cyber attacks	cyber actors and actions	cyber generic	cyber protection
CYBER NIST	malware cipher block chaining backdoor	hacker attacks	audit audit log access access control list	cryptography authentication digital signature identiy proofing
CYBER ISO	spam vulnerabilities residual risk	thrreat	audit electronic infrastructure informative system	cybersecurity authentication cryptography user identification data antivirus risk assessment
CYBER WHITEBOOK	infected pc malicious code injection exploit spam	attack attempt hacking cyber criminals attack techniques	operative system digital instrastructure account cyber space	strategic supervision asymmetric cryptography incident response informative system security cybersecurity digital security privacy
CYBER GLOSSARY	cyber threats spoofing espionage	threat risk	critical infrastructure	cybersecurity authentication cryptography encryption
CYBER CLUSIT	virus espionage malicious code vulnerabilities data breach	attack attempts hacking threat	remote access cloud e-mail	technology protection inciden response data integrity cybersecurity protection pricacy cryptography

Figure 4.3: Synthesis of the main first candidate terms with reference to the shared knowledge contained in the evaluation lists

BERT is based on the assumption that terms are connected with their contexts and, by consequence, to next sentence prediction, the first sentence given to BERT is the one which contains the term, and the sentence to predict is the term itself. In this case, for the research objectives, the training run for BERT has been centered on training this model with all context/term pairs that are present within the source corpus conceiving them as positive examples, whilst randomly generating the negative ones. BERT performed efficiently with the Italian heterogeneous corpus of Cybersecurity, giving back a 1568 term list from the extraction and new single and complex terms to be considered as part of the semantic network, e.g., *advanced fraud protection, advanced persistent threat, alphastar, altcoins, anti-botnet, anti-ddos, banking trojan backswap, bitpay, cracking ddos, cyber-ricattatori, darktrace, eternal petya, grayhat, kessem backswap malware, mobile malware, phishtank, weaponization*, etc. Apart from BERT, the term extraction systems cover their main functionalities key filtering techniques which can enable users to fix customized thresholds to be applied according to different statistical measures above which the ranked candidate terms are kept. To guarantee a correct implementation of statistical measurements to compute Recall, no additional filtering techniques have been applied except for the one already contained as as default parameters.

Table 4.7 gathers the total number of extracted candidate terms for each tool/method, T2K and BERT software outputs, are respectively the largest and the smallest terminological set. The scores highlight the higher level of Recall especially related to the extractions run with PKE library and T2K linguistic-oriented software. The low total number in the Precision measurements is clarified by the fact that the size of the evaluation is not particularly high (around 200) and the systems output is often around thousands of terms. Furthermore, the evaluation lists, i.e., the gold standards terminology frameworks, are not exhaustive and, as a result, do not provide a fair evaluation on Precision. A range of correct terms, for instance, not present in the evaluation lists have been observed. Finally, based on the F1 score, BERT obtained the best scores proving that even without specific filtering applied it can efficiently elaborate specific types of semantic outputs.

Table 4.7: T2K, TermSuite, PKE, and BERT term extractions executed on Cybersecurity corpus. The evaluation is performed over five lists (Clusit, Glossary, Nist, Iso, and Cyber) and the results (%) are provided with respect to Precision (P), Recall (R), and F-measure (F1)

Corpus	coverage (%)	Tools	Evaluation lists												
			Clusit 61.3			Glossary 72.5			Nist 35.3			Iso 67.0			Cyber 100		
			P	R	F1	P	R	F1	P	R	F1	P	R	F1	P	R	F1
CyberSec		T2K	0.01	23.7	0.02	0.02	42.2	0.04	0.05	21.2	0.10	0.01	47.7	0.02	0.01	30.7	0.02
		TermSuite	0.10	7.92	0.19	0.37	21.8	0.73	0.77	9.98	1.41	0.12	23.8	0.25	0.20	13.7	0.40
		PKE	0.05	**49.0**	0.10	0.06	**44.0**	0.12	0.12	**21.2**	0.24	0.02	44.3	0.04	0.05	**46.5**	0.10
		BERT	**0.48**	14.3	**0.93**	**1.04**	15.1	**1.95**	**2.30**	7.10	**3.47**	**0.43**	20.4	**0.84**	**1.11**	25.9	**2.13**

4.2.3 Expert support system

The main objective of this present study is to enhance the current structure of the thesaurus, developed by using semi-automatic term extraction techniques, to ensure a representative semantic monitoring tool covering the information of a specialized domain through a fine-grained network of semantic relations configuration. This achievement has been reached out also by exploiting the collaboration with the group of experts of the Cybersecurity domain, who played a crucial role in the validation process of the semantic first data model in terms of accuracy of candidate terms relying on their expertise and specialized skills by taking as input the output of the term list obtained by the mapping structure. In particular, the group of experts consulted for the realization of the two semantic means of the Cybersecurity terminology control is represented by the IIT-CNR Institution, particularly referred to the CyberLab team that developed the OCS web portal where the thesaurus and the ontology are inserted as services providing a guided semantic knowledge about the domain of study.

Validation from experts resulted fundamental both for what concerned the approval or rejection of certain terms that had been put in primary positions by frequency measurements algorithm in the output lists of the term extractors, and for the additional semantic information granted from their expertise that enhanced the data framework of the thesaurus. The consensus of the experts (Schultz, 1968) enables the composition of the first beta version of the thesaurus, as well as, starting from this model, the development of the taxonomy structure in the OWL language. The following Figure 4.4 shows some of the new entry terms the collaboration with the qualified group of professionals in the domain of Cybersecurity has imported with respect to the mapping architecture obtained by associating the term extraction lists with the gold standards taxonomies. As it can be observed, there are some terms in red relating to the main malware names the experts provided starting from the semantic grouping system offered by the term extraction lists, that have been found in the evaluation list in a second time.

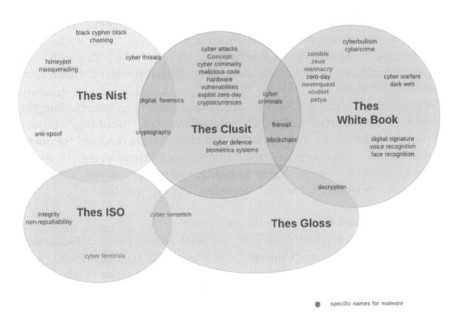

Figure 4.4: New terms derived by the collaboration with the group of experts

As done for the term extractions mapped with the gold standards, also the terms within the thesaurus, *Thes*, selected after having undergone a validation phase carried out with the Cybersecurity experts, have been mapped with the aforementioned gold standards, and the following show some example:

- **Thesaurus and NIST** 'access control,' 'Access Control List (ACL),' 'anti-spoof,' 'audit,' 'audit log,' 'back door,' 'block cipher,' 'cyber incidenti' (*cyber incidents*), 'cyber terrorismo,' 'cyber terroristi,' 'digital forensics,' 'honey-pot,' 'malware,' 'masquerading,' 'cipher block chaining,' 'random bit generator (RGB),' 'digital signature,' 'identity proofing/'

- **Thesaurus and ISO**: 'audit,' 'autenticazione' (*authentication*), 'confidenzialità' (*confidentiality*), 'controllo degli accessi' (*access control*), 'cyber terrorismo' (*cyber terrorism*), 'cyber terroristi' (*cyber terrorists*), 'disponibilità' (*availability*), 'integrità' (*integrity*), **'non ripudiabilità'** (*non-repudiability*), 'rischio residuo' (*residual risk*), 'sicurezza informatica' (*cybersecurity*), 'vulnerabilità' (*vulnerabilities*), 'valutazione del rischio' (*risk evaluation*).

Thesaurus and White Book: 'access control,' 'account,' 'audit' 'autenticazione' (*authentication*), 'biometria' (*biometrics*), 'bitcoin,' 'blockchain,' 'botnet,' 'cifratura' (*encryption*), 'commercio elettronico' (*e-commerce*), 'conficker,' 'confidenzialità' (*confidentiality*), 'criptovalute' (*cryptocurrency*), 'crittografia' (*cryptography*), 'cyber crime,' 'cyber spazio' (*cyber space*), 'cyber warfare,' 'cyber warriors,' 'cyberbullismo'(*cyber bullism*), 'cybersecurity,' 'dark web,' 'data breach,' 'decifratura'(*decryption*), 'deep web,' 'digital forensics,' **'dridex,'** 'e-commerce,' **'ethereum,'** 'exploit,' 'firewall,' 'firma digitale'(*digital signature*), 'hacking,' 'hash,' 'home banking,' 'impronta digitale'(*fingerprint*), 'integrità,' 'malware,' **'neverquest,' 'nivdort,'** 'password,'**'petya,' 'phishing,'** 'privacy,' 'profilazione'(*profiling*), **'ransomware,' 'riconoscimento facciale,' 'riconoscimento vocale,'** 'rischio residuo,' 'riservatezza,' **'social engineering,'** 'software,' 'software antivirus,' 'spam,' **'spear phishing,'** 'spoofing,' 'token,' 'virus,' 'vulnerabilities,' **'wannacry,' 'worm,' 'zero-day,' 'zeus,'** 'zombie/'

Thesaurus and Clusit 'attacchi cibernetici' (*cyber attacks*), 'attacchi mirati' (*targeted attacks*), 'autenticazione' (*authentication*), 'bitcoin,' **'blockchain' 'botnet,' 'code** injection,' **'Concept,'** 'confidenzialità' (*confidentiality*), 'criminali informatici' (*cyber criminals*), 'criminalità informatica' (*cyber criminality*), 'criptovalute' (*cryptocurrences*), 'crittografia' (*cryptography*), 'cyber criminali' (*cyber criminals*), 'cyber defence,' 'cyber forensics,' 'cyber terrorismo' (*cyber terrorim*), 'cyber terroristi' (*cyber terrorists*), 'cyber-hygiene,' 'cybersecurity,' 'dark web,' 'data breach,' 'digital forensics,' ' disponibilità' (*availability*), 'e-commerce,' **'exploit zero-day,'** 'firewall,' 'firma digitale' (*digital signature*), 'furto di dati'(*data breach*), 'hacking,' 'honeypot,' 'incidente informatico' (*cyber incident*), 'malware,' 'intercettazioni' (*interception*), 'password,' 'phishing,' 'ransomware,' 'rischio residuo' (*residual risk*), **'riservatezza'** (*privacy*),'sicurezza informatica' (*cybersecurity*),'sistemi biometrici' (*biometric systems*), 'software malevoli' (*malicious software*), 'spam,' **'token,'** 'virus,' 'vulnerabilities,' **'vulnerabilità hardware'** (*hardware vulnerabilities*), **'wannacry.'**

Thesaurus and Glossary of Intelligence 'autenticazione'(*authentication*), 'cifratura' (*encryption*), 'crittografia' (*cryptography*), 'cyber defence,' 'cyber terrorismo' (*cyber terrorism*), 'cyber terroristi' (*cyber terrorists*), 'cybersecurity,' 'decifratura' (*decryption*), 'decrittazione,' 'integrità' (*integrity*), 'sicurezza informatica' (*cybersecurity*), **'spoofing,' 'aise,' 'aisi.'**

4.3 Automatization of thesaurus construction

The methodologies concerning the automatization of the thesauri configuration structure [50, 57] are focused on the discovery of a technique that can facilitate the management of semantic relations networks that are meant to systematize the specialized knowledge-domains. The achievement of an automatized procedure implies a set of approaches to be tested that range from the configuration of patterns oriented recursive structures detection (Condamines, 2007), or following some experiments present in the literature such as that proposed by Grefensette (1994) about the automatization of thesauri, or along the same line Hasan's work (2014) [98] who studied the common entries in two different thesauri by building up pairs of codes.

Semantic similarity approaches have been in a first place employed for the detection of in the single word terms (SWTs). The procedures that have been proposed by many authors in the process of recognizing recursively frequent portion of texts are based on the lexicon [174], on the multilingualism frame [77, 111] [182], the distributional [116], Daille (2014) and distributed semantics Mikolov (2013), Bojanowski (2016).

4.3.1 Approaches

In the following sections three approaches tested for the relationships identification automatization to be imported in the thesaurus are described. Specifically, the first one is referred to the **variants recognition** system provided by the TermSuite terminology extractions; the second one covers the tasks implemented in the **pattern-based configurations** aiming at retrieving the hierarchical, synonymous, and associative recursive structures inside the source corpus segmenting the three semantic connections with the respective key verbs, noun phrases, adjectival chains, and prepositional constructs; the third method is based on the **word embedding models** execution over the textual documentation of the corpus through word2vec and fastText alogorithms, giving for each semantic relation a range of examples that explain the connection among terms and the enhanced structure applied to the beta version of the Italian Cybersecurity thesaurus.

4.3.2   Variants

TermSuite toolbox presents a key function related to the semantic variants detection over large source corpora. It executes syntactic and morphological predefined rules written in JAVA according to the language over a set of textual documents given in input, and implements them in the textual processing operations that give as output a list of terms sorted according to their variants associations, as described in greater details in the section 3.4.3.1 and in the previous section 4.1.2.2. This procedure helped in identifying in a more precise way some of the hierarchical and synonymous structures inside the corpus to be included as candidate entry terms in the thesaurus.

4.3.3   Patterns-based

One of the aims of this work is the discovery of a method able to activate inference processes with respect to the semantic relations network discovery to be included within the candidate terms of the Italian Cybersecurity thesaurus. Therefore, to retrieve the semantic connections starting from a domain-oriented data set, patterns-based configurations have been exploited as one of these recognition systems (Roesiger, 2016). For this purpose, key verbs have been selected in order to gather the main connecting patterns to represent the causative, hierarchical, synonymous and associative relationships among the terms included the Italian Cybersecurity source corpus texts. The objective of this task is to automatize the entangled network of semantic connections that characterizes the thesaurus outline and to isolate the most useful recursive expressions to be then used as Object Properties in the ontology structure.

The first verbs addressed to agent - cause correlations aimed at improving the level of associative relationships accuracy, proper to thesauri network system and labelled as RT, *Related Terms* (ISO 25964-1:2011). Indeed, Roesiger (2016) analyzed the ways through which to achieve efficient datasets reflecting the semantic relationships following three principles:
- the employment of NLP techniques;
- the selection of specific verb-object pairs that deal with the fields of knowledge pertinence and relevance;
- the assumption that patterns can be syntactically correct.

The selection started for the causative patterns configuration has not followed frequency criteria, rather the majority of them represent the most used syntactic structures that express the cause associations, as well as hyperonymy and meronymy constructs and synonymy sets, in Italian language, with the objective of retrieving the co-occurrences in the source corpus covering these morphosyntactic chains. As previously mentioned, using patterns, especially referred to the causative relations (Lefeuvre, 2015), points at guaranteeing an improved systematization of the related terms structure fixed in the Cybersecurity thesaurus, which sometimes proves to be insufficiently descriptive for the domain concepts correlations. In fact, the associative relationship in thesauri, differently from the hierarchical and equivalence connection, is the one showing more ambiguity in interrelating the domain-oriented terms with each other, since it misses a more detailed form of semantic description. Hence, the employment of causative-based patterns enables the creation of semantic inter-references from one specific term to another in a more precise and reliable way, always taking into account as authoritative sources of information the documents present in the compiled source corpus. The lists that follow provide examples for what concerns, firstly, the selected causative verbs, and successively the hierarchical and equivalence relations. Among the new discovered forms of connections in terms present in the source corpus, new information about the correspondences among the Cybersecurity specialized terms has been recognized, as for example the relation between *camouflage* and *password*, or *cyber threats* and the *security properties*; sometimes the new associations confirmed the pre-configured structure of the thesaurus semantic network, as, for instance, *cyber attacks* and *DDoS* or *spoofing*.

The causative connections found in the source corpus by the means of patterns-guided configurations, provided extra semantic information to be adapted to the pre-determined set of connections in the Italian Cybersecurity thesaurus, which has already undergone a validation process by the group of experts of the domain, i.e., the CyberLab teamwork. The following Figure 4.5 shows some cases where the related terms, RT, in the thesaurus proved not to be completely adequate to exhaustively represent a description of the candidate terms' relations with others in the domain – even though from a qualitative point of view the thesaurus relations are suitable for the purposes of managing the domain-specific information – and, on the other hand, the way the patterns-pair sets have supported the configuration of a more precise semantic network.

4.3.3.1 *Causative relations*

The main verbs, mostly in the passive form, representing the causative relations, that is a connection that links together an agent that provokes a determined situation (Roesiger, 2016), are found in the more general-language oriented and in the domain-targeted ones and have been picked randomly from available Italian verbs classification:

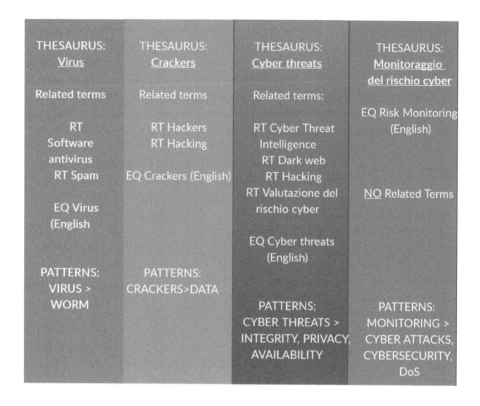

Figure 4.5: Comparison of RT in thesaurus and patterns paths outputs

- **general-language:**
  alterare (*alterate*)
  causare (*cause*)
  condurre a (*lead to*)
  danneggiare (*damage*)
  forzare (*force*)
  impattare (*impact*)
  impedire (*prevent*)
  indurre (*prompt*)
  influenzare (*influence*)
  innescare (*trigger*)
  ostruire (*obstruct*)
  pesare su (*burden to*)
  pregiudicare (*preclude*)
  provocare (*provoke*)
  rallentare (*decelerate*)

suscitare (*raise*)

tradursi in (*translated in*)

■ **domain specific**:

cifrato da (*encrypted by*)

contaminato da (*contaminated by*)

infettato da (*infected by*)

monitorato da (*monitored by*)

proteggere da (*protected by*)

tramettere da (*transmitted to*)

aggirare (*bypass*)

attivare (*activate*)

catturare (*capture*)

codificare (*encode*)

sfruttare (*exploit*)

colpire (*attack*)

sfruttare (*exploit*)

veicolare (*vehiculate*)

rubare (*steal*)

manomettere (*sabotage*)

comportare (*imply*)

The following are some of the main examples for the domain-specific causative verbs that enhanced the structure of the thesaurus:

■ **provocare (*to provoke*)**:
> virus – worm
> cyber attacks – DDoS
> risks – cyber threats

■ **danneggiare (*to damage*)**:
> crackers – data
> cyber threats – integrity, privacy, availability

■ **comportare (*to imply*)**:
> cyber attacks – malware
> cyber harassment – cyber bullism

■ **alterare (*to alter*)**:
> camouflage – password
> spoofing – cyber attacks

■ **manomesso da (*sabotaged by*)**:
> monitoring – cyber attacks
> monitoring – DoS

■ **impattare (*to impact*)**:
> DDoS – cyber attacks
> monitoring – cybersecurity

4.3.3.2 Hierarchy

Hyperonymy and Hyponymy

The patterns constructs hereafter summarized are the ones that represent the hierarchical connection, sometimes there are alternatives marked with the symbol '/,' and other times where an item in the recursive pairs is optional presents the question mark as its symbol:

- caratterizzato da/come (*characterized by/as*)
- come ad esempio: (*as for example:*)
- composto da (*composed by*)
- consistere in (*consisting in*)
- contrassengato da (*labeled as*)
- costituito da+NOM?+ (*constituted of + NOM*)
- di tipo (*type of*)
- come (*as/like*)
- è un (*is a*)
- ""
- il più +ADJ (*the most+ adj*)
- in particolare+PRON (*in particular + PRON*)
- NOM + definito come (*NOM + defined as*)
- NOM+più+ADV (*NOM+more+ADV*)
- rappresentato da (*represented by*)
- specificato da/come (*specified by/as*)
- tracciare da (*traced from*)
- tra tutti è (*among everything is*)
- è un tipo di (*is a type of*)

Meronymy
- essere membro/parte/elemento di (*being member/part/element of*)
- essere tra PRON/DET (*being among the*)
- essere uno dei (*being one of*)
- fare parte di (*being part of*)
- forma di (*form of*)
- (NOM) + include + NOM :? (*NOM+include+NOM*)
- NOM +comprendere+ NOM (*NOM+comprehend+NOM*)
- tra cui(*among which*)
- :
- (

4.3.3.3 Synonyms

A short list of the main configuration patterns to identify the synonyms in the source corpus is the following:

- , conosciuto come (*, known as*)
- essere visto come (*be seen as*)
- essere chiamato (*be called*)
- essere simile a (*be similar to*)
- vale a dire (*that is*)
- denominato con il nome (*denominated with the name*)
- avere in comune con (*have in common with*)
- noto con il nome di (*famous with the name of*).

4.3.4 Word embeddings detection

Two word embedding models, Word2Vec (W2V) (Mikolov, 2013b) and FastText (Bojanowski, 2016), have been executed over the source pre-defined semantic relations in the thesaurus, which had already passed the approval check by the domain experts, to automatize the identification of the semantic relationships to be managed in the thesaural configuration. For both models the Skipgram (Sg) and the Continuous Bag of Words (CBOW) models have been tested and the final scores are given in terms of Precision at 100 (P@100), each MWT is considered as a sum of its constituents embeddings (Arora, 2017, Hazem, 2018).

Table 4.8: Output of the semantic connections extractions run with word2vec (W2V) and fastText models using the Precision at 100 (P@100%) score.

	Hyp	Syn	Rel	Cause
W2V (Sg)	**5.91**	**45.7**	5.38	**13.2**
W2V (CBOW)	2.95	34.2	6.15	0.00
fastText (Sg)	4.73	34.2	**10.3**	3.77
fastText (CBOW)	3.55	22.8	**10.3**	1.88

The best scores achieved by implementing the two word embedding models are referred to the synonyms detection [17] with Sg (45.7%). The reason why the automatic detection of the other types of semantic relations present a lower score is referred to the mixed nature of language in the source corpus terminology. Indeed, this specialized domain is steeped in the borrowings from English implying a heterogeneous characterization of the documentation from which terms come from, and this might undermine the embedding models with respect to low frequency terms, but not on the qualitative level, as it can be further observed.

Word embedding models helped in retrieving the proximal terms starting from pre-configured connections present in the first beta version of the Italian Cybersecurity thesaurus. Therefore, the distributional similarity granted by the proximity processing tasks is analalyzed using the hierarchies, synonyms pairs and related structures. These models supported the validation phase of the already existing semantic relations among specialized terms and the integration process referred to the enhancement of the associative relations, which, as previously underlined, result to be less descriptive in the organization and representation of the domain terms. As the examples below demonstrate, some of the terms retrieved to be very close to the relation searched up in the execution of the word embedding models confirmed the existing kind of semantic interrelation (for instance, broader term with their more specific ones, or synonyms, have been verified), but in the same proximal groupings it has been possible to find other distributed terms that enhanced, in turn, the associative relation in terms of precision with respect to previously determined configuration in the thesaurus. As a result, in the hierarchical pairs, synonyms as well as causative detection, it has been possible to find other related terms to be integrated to the one present in the thesaurus approved by experts configuration.

The following are some of the main outputs obtained by executing the word embedding models over the already configured and approved by the domain experts semantic relationships within the Cybersecurity Italian thesaurus.

Hyperonyms detection

1. **gestione del rischio cyber (*risk management*)** which has as hyponym *piano di risposta al rischio cyber (risk response measures)*, has been connected with: *attacchi cibernetici (cyber attacks), cavalli di troia (trojan horse), cyber intelligence, difesa informatica (cyber security)*; this confirms the thesaurus outline regarding the top term category of *cybersecurity* and adds another one to be considered, i.e., *cyber intelligence*.

2. **intrusion detection system** – host-based, in the thesaurus is the hyponym of *network security systems*. Among the terms related in a hierarchical way, *network security systems* has been confirmed, and, in turn, other related terms have been included in the semantic structure as RT, e.g., *hacker, mid hacking, sniffing* and *malware*.

3. **chiavi crittografiche (*cryptography keys*)** has as hyponyms *chiavi crittografiche private (private cryptography keys)*, in the models another broader term retrieved is *tecnologie crittografiche (crypthograpy technologies*, normalized in the thesaurus as *cryptography*. The word embedding implementation shows the possibility to add as relative terms for their proximity the following ones:
   RT: *anti spam, trojan horses, cognitive computing, hacking, internal threats, personal data confidentiality and systems vulnerabilities*.

4. **trojan horse** which has as its specific term *nivdort*, has been found, confirming the structure of the thesaurus, with the hyperonym *malware* and

added the following related terms:
RT: *anti spam, attackers, cyber defence, hacking, cyber risk, risk management and cryptography technologies.*

5. **crypto miner malware** with its hyponym *bitcoin virus* confirms the hyperonyms *attack, cyber attacks*, which in the thesaurus are presented in the form of *cyber attacks mechanisms* and *malware*. The models integrated the related terms, among these only *trojan horse* was present in the thesaurus:
RT: *anti spam, cyber defence, virus spread, hacking, netbus, cyber risk, personal data confidentiality, risk management, e-mail software and cryptography technologies.*

6. **cyber minacce (*cyber threats*)** hyperonym of cyber bulli (*cyber bullies*) is found, validating the thesaurus outline, with broader terms *cyber attacks, cibernetic attacks, attack*, and adds to the existing related terms in the thesaurus (*cyber Threat Intelligence; dark web; hacking; risk assessment*):
RT: *anti spam, trojan horse, deep security, cyber defence, cyber risk and cryptography technologies.*

7. **cyberbullismo (*cyberbullsim*)** – cyber gang confirmed the hyperonym *cyber attacks* and added these associative connections: RT: *trojan horse, malware, disciplinary actions, kids, cyber risk and risk management*
while showing some discrepancies with the semantic meaning in the association with *spam* or *cryptography technologies.*

8. **cybersecurity** which contains as more specific unit *meccanismi di attacchi informatici (cyber attacks mechanisms)* confirmed the narrower term *trojan horse*, added the broader one *materia di criminalità informatica (cyber criminality)* and the following related: RT: *blockchain, hackering, cyber risk, risk management.*

9. **dos** present *ddos* as its narrower term, and in the models it is found together with the broader term *attack*, which validates the structure agreed upon with the group of experts. The models enhanced the association concatenation by integrating it with:
RT: *anti spam, trojan horse, cyber defence, terrorists, computer emergency response team, intellectual properties, mac hacK, risk management and terrorist*
putting also some verbs next to the chains: *defeat* and *encrypted.*

10. **furto di informazioni personali (*personal data breach*)** presents as its narrower term *spoofing* and the models confirmed the hierarchy with *cyber attacks*, adding the following related terms:
RT: *anti spam, data storage, trojan horse, cyber defence, intellectual properties, cyber risk, personal data confidentiality, risk management and cryptography technologies.*

11. **meccanismi di attacchi informatici (*cyber attacks mechanisms*)** together with its narrower term *back door* is identified by the models with its more generic term *cyber attacks* – present in the thesaurus as synonym –, *cyber criminality*, and *internal threats*. The associative integrations cover: RT: *anti spam, cyber defence, hacking, incidents monitoring, e-payments, possible threats, cyber risk and cryptography technologies*.

12. **meccanismi di attacchi informatici (*cyber attacks mechanisms*)** with its hyponym *social engineering* confirmed the hierarchical structure with the more generic *cyber attacks*, then added *internal threats*. For what regards the related terms it is possible to find extra information: RT anti spam, mail boxes, cyber defence and hacking.

Synonymys detection

1. **cybersecurity** has been related to the following synonyms that can be considered as positive candidates for the thesaurus: *difesa informatica (cyber defence), deep security, sicurezza cibernetica (cibernetic security), protezione cibernetica (cibernetic protection), sicurezza dei sistemi informativi (informative systems security) and sicurezza ict (ict security)*.

2. **malware** has been found related with these synonyms: *software malevolo (malicious software), programmi malevoli (malicious programs)*, confirming the existing synonymous structure in the thesaurus; the interesting result is that *malware* is associated in the same list with several representative terms that are conceived as candidates to improve its semantic associative connections: *spyware, keylogger, firewall and exploit*.

3. **cyber incidents** confirmed the synonyms fixed in the thesaurus with *cybernetic incidents* and *digital incidents*, and, also, integrated the related connections:
   RT: *anti spam, authentication , cryptography, cyber defence, crisis management, hacking, security mechanisms, incidents monitoring, cyber risk, informative systems security, e-mail software, cryptography technologies.*
   Word embedding models also gave as narrower term *trojan horse* and hierarchy with *cyber criminality* while in the thesaurus it is established as *cyber attacks mechanisms*

4. **software malevoli (*malicious sofware*)** confirmed the synonym *malware* and integrated the related terms:
   RT: *anti spam, hacking, cyber risk, e-mail software and cryptography technologies*
   The models included also some more specific terms as *backdoor* confirming *trojan horse*.

Related terms detection

1. **zero-day** that in the thesaurus is connected on an associative level with *software vulnerabilities*, is grouped together with *trojan horse, anti spam, hacking, privacy and risk management.*

2. **cyber molestie (*cyber harassment*)**, related in the thesaurus, among others, with *cyber stalking*, has improved structuring matches since it is found associated also with *cyber theft, hacking threats and cyber insurance.*

3. **cybersecurity** related to cybercriminalità (*cyber criminality*) is found with:
RT: *blockchain, trojan horse, cyber defence, cyber-hygiene, e-payments, cyber risk and risk management.*

4. **cyber bulli** is a related term in the thesaurus of *cyberbullismo(cyber bullism)*, the models found it also related with:
RT: *computer emergency response team, cyber-hygiene, cyber risk, risk management, warfare, attack and anti spam.*

5. **cyber stalking** related with *dark web* is also retrieved with associative connections with the following terms:
RT: *anti spam, attacks, trojan horse, cyber defence, security leaks, cyber criminality, mid hacking, network address translation, cyber risk, risks and incidents, risk management, security and defence, sniffing, warfare and web content filter*
The word embedding models also highlighted some connecting verbs expressions, such as *prevenire una minaccia (to prevent a threat).*

Causative relations detection

1. **cyber minacce (*cyber threats*)** was connected through the causative verb *to damage* to the *security properties of data*, in these models it is linked to *cyber intelligence, difesa informatica (cyber security), hackeraggio (hacking) and cavalli di troia (trojan horse).*

2. **bitcoin** was associated with *data loss* through the causative pattern verb *to prevent*, with the application of these embedding techniques it also seems related with *risk management, cyber risk, spam and hacking.*

3. **attacco (*attack*)** was connected through the causative verb *(to imply)* to *malware*, the models added:
anti spam, attack, brute force, trojan horse (in the thesaurus is its NT), *cyberattack* (acting in the thesaurus as its synonym), *cyber defence, hacking, hard hacking, cyber risk , risk management, authentication system, sniffing, cryptography technologies, terrorists, fraudulent usage and worm* (NT in the thesaurus).

4. **dati (*data*)** are found connected with *hacker* through the causative verb *to alterate*, the models integrated:
   data cryptography, digital data, e-payment, cyber risk, personal data confidentiality and e-mail software.

5. **spoofing** is linked to *cyber attacks* with the causative verb *to alterate*, here the word embeddings added the causative verb *to decode* and enhanced the related structure with the following connections: *anti spam, attack* (its broader term), *trojan horse, cyber defence, hacking, internet service provider, koobface, login, malware* (BT in the thesaurus), *automatic recognition and cyber risk.*

Though the relevant contribution the word embeddings models provided in enhancing the associative semantic relationship of the thesaurus with more precise techniques of semantic proximity detection that guaranteed the accuracy evaluation process of terms selection in the thesaurus structure, it is also true that sometimes the outputs resulted rather similar in the occurrences they provided and, at times, not very faithful, e.g., *cyber gang* is connected in an hierarchical way with *cryptography*. On the other hand, the synonyms detection showed better results and the findings are very exhaustive both for what concerns the retrieval of the synonyms themselves, and for the recognition, among the outputs given by the models, of other candidate related terms to add in the thesaurus.

The connections given by these models were performed using the existing thesaurus relations, which have been created following the ISO 25964-1:2011, 25964-2:2013 rules, as source correspondences to be enhanced with sophisticated grouping procedures. Though the manual evaluation of these series of interrelations has inferred quite similar proximities among the terms extracted in all the four classes of relations, at least on a quantitative level, e.g., *rischi cyber (cyber risks), anti spam, hackeraggio (hacking)* appear for almost all the cases, many associated terms helped in improving the thesaural systematization. It should be underlined that when evaluating these kinds of lists, a minimum level of knowledge expertise about the technical domain to be studied is required since many terms connected with the head ones sometimes appear related in a very implied way, at least for the domain experts, e.g., *cavalli di troia (trojan horse)* or *zero-day*.

Chapter 5

Semantic tools for Cybersecurity

In this chapter the thesaurus and the ontology semantic tools managing and representing the domain of Cybersecurity will be presented. In the first section the thesaurus will be described under the twofold level of applicability, that is the integration in the OCS web portal implemented with the automatized techniques execution for the semantic relationships retrieval. Specifically, the macro categories, the relations of hierarchy, equivalence and association will be presented. The aim of this description is to provide a overall outline of how the thesaurus can support and orientate the understanding of specialized fields of knowledge allowing to systematize in a semantic connected structure their technical terminologies. Successively, the sections will cover the steps addressed to the reengineering process of the terms contained in the thesaurus to an ontology systematization made up of classes organized in a taxonomic distribution and linked together with OWL Object and Data Properties. This chapter ends with an highlight of the main differences that exist in the configuration of the thesaurus' associative relationship, a connection representing the relations occurring among the domain-specific classes that the OWL syntax expresses in a more explicit way.

5.1 Construction of Italian Cybersecurity thesaurus

The Italian Cybersecurity thesaurus has been developed with the tool Tematres – Controlled Vocabulary Server[1], and, as previously described in Chapter 4, the

---

[1]https://www.vocabularyserver.com/

realization of this semantic tool has implied a series of steps towards its final configuration, which can be summarized as follows:

1. reference to the existing state of the art for the identification of the equivalent semantic tools to be classified with respect to the same field of knowledge, and a census of what are the domain-specific gold standards with which the mapping process can be run to start checking the semantic coverage threshold of the source corpus in relation to the official shared knowledge-domain;

2. selection of authoritative portals in which the source documents for the population of the corpora (ita and en) are meant to be collected from;

3. definition of the trustworthiness criteria according to which a corpus can be defined as reliable and representative, e.g., the time range, the language, the sources, the applicability to the domain, the heterogeneity of the texts to cover the terminology as much as possible form a technical perspective;

4. analysis of the compiled corpus;

5. extraction of terminology by using pre-trained software or algorithms working with NLP;

6. selection of the candidate terms by means of frequency approaches and specificity distribution studies;

7. support from experts in the steps concerning term validation and integration within the thesaural structure thanks to their expertise and skills in the domain background;

8. realization of the first beta version of the semantic tool for specialized domain meant to be automatized;

Integration of the methodologies presented in this research work – Techniques towards the terminology enhancement coverage[2]

9. applying intersection measures between corpora and reference texts, i.e., mapping tasks with the gold standards referring to the domain under analysis, and with the support of a group of domain experts that guide the filtering process of the preferred terms identification (which have eventually already been grouped with respect to the aforementioned criteria) to be inserted in a structured organization of specialized terminology;

---

[2]from point 9 a description of the integrations provided by this study activities is given

10. automatization of the semantic relationships to be configured in the thesaurus' knowledge organization structure, i.e., *hierarchical, synonyms, associative* connections, through variation detection, pattern-based queries to be run over the source pos-tagged source corpus documentation, word embedding models to reveal the proximity among terms starting from the pre-established relations inside the thesaurus;

11. terms and semantic network contained in the thesaurus converted into an ontology system to support the better form of explication missing in the associative connection proper to thesauri through Object Properties written in RDF and OWL syntax;

12. multilingual alignment with English;

13. monitoring the terminology variation through the updates of the source documents through SPARQL queries and the perspective of following up on changes coming from authoritative profiles on social media contexts.

The phase covering the validation of experts supported the decision-making process of the main macro-categories, which become in the thesaurus syntax *Top Terms* (TT) to be selected in order to create the first entries of the semantic monitoring tool. Since the rules behind the hierarchy relation meant to be placed in a semantic resource indicate that terms are to be structured following the super-ordinate and subordinate principles, the agreement with experts moved towards a categorized outline in which some of the more specific terms appearing with the main areas are defined as *Subject categories* marked as SC, such as *Cyber threats* or *Cyber terrorism*, which, in turn, are also conceived as Broader terms, BT for other candidates; for instance the tree-shaped structure

Cyber criminality (TT) → Cyber threats (SC) → Cyber threats (BT)

is twofold since *threats* include other more specific terms, NT, as the following Figure 5.1 shows

The abstraction process towards the definition of these super-ordinate grouping headings is focused on the following structure:

■ Cybersecurity: Cyber threats, Data, Risk management, Cyber attacks mechanisms, Cybersecurity mechanisms, Vulnerabilities;

■ Cyber Defence: Cyber war, Cyber terrorism, Cyber-hygiene, Interceptions;

■ Cyber Bullism: Cyber gang, Cyber harassment, Cyber stalking;

■ Cyber criminality: Cryptocurrences, Cyber laundering, Cyber threats, Cyber terrorism, Cyber espionage.

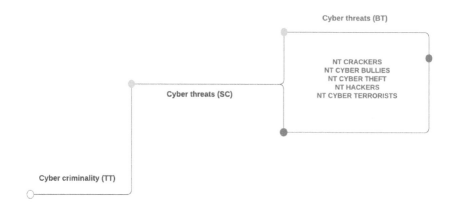

Figure 5.1: Subject category and broader term

Once these macro-areas have been selected under the experts' supervision, the thesaurus semantic configuration started by adding its main basic semantic relationships of hierarchy, equivalence and association. Being the main objective of this research joint collaboration project with IIT-CNR the realization of a terminology monitoring and organizing tool for the domain of Cybersecurity specifically in Italian language, the thesaurus is in its first conformation provided in Italian and the notes, SN, that come together with terms, where needed, have been translated from the main English gold standard repositories into Italian language too. The multilingual alignment has come further in order to give a bilingual resource containing term equivalents. In the sections below an overview of the semantic network the thesaurus is able to guarantee for the technical terminology proper to specialized fields of study is provided, starting with the hierarchical diagram for what regards the terminology connections for the *Cybersecurity* subject category, moving the examination towards the synonyms contained in the thesaurus and giving tables for what regards the associations given in the controlled vocabulary and the notes derived from the authoritative documents within the corpus.

5.1.1 Semantic relationships

5.1.1.1 Hierarchy

Diagram 5.2 depicts the hierarchical structure realized for the subject category *Cybersecurity*. The major sub-subject categories, *Cyber threats, Cyber risk management, Cyber attacks mechanisms, Cybesercurity mechanisms, Security Properties, Vulnerabilities* are considered as the broader terms of the arch generated tables which included the more specific terms, which, in turn, can become broader terms of some specific ones as in the case of

Cyber attack mechanisms [BT] → *Spam [NT]* / *[BT]* → *Phishing [NT]* / *[BT]* → *Spear Phishing Smishing*

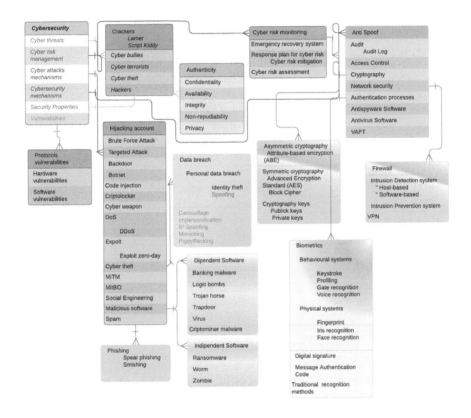

Figure 5.2: Cybersecurity hierarchical structure

5.1.1.2   Equivalence

Figure 5.3 below enumerates the synonyms included in the thesaural structure, the terms on the left-hand-side represent the preferred lexical units, which have mostly been agreed upon with the support of experts who attributed a more frequent use of some of the terms instead of others – this is the case of some of the acronym forms, such as *DDos* – that have been selected as main entry forms

preferred term	not preferred term
Advanced Encryption Standard	AES
Attribute-based encryption	ABE
Crackers	Digital pirates, Black hats
Cryptography	Encryption
Cyber forensics	Digital forensics
Cyber guerra	Cyber war
Cyber incidents	Cibernetic incidents, Digital incidents
Cyber threats	Cyber Threat Actors
Cybersecurity	Informative systems security, ICT security, Cibernetic security, Cyber protection, Deep security
Dark web	Deep web
DDoS	Distributed Denial of Service
DoS	Denial of service
Hackers	Cyber criminals, White hats
Cyber attacks mechanisms	Cyber attacks, Cibernetic attacks, Attacks
Message Authentication Code	MAC
MitM	Man-in-the-middle Attack
SCAP	Security Content Automation Protocol
Malicious software	Malware
Spoofing	Sniffing
Cyber risk assessment	Cyber risk analysis
VAPT	Vulnerability Assessment and Penetration Testing
VPN	Virtual Private Network

Figure 5.3: Equivalent terms

because of their persistent domain-specific usage, which are provided on the right-hand-side of the table.

5.1.1.3 Association

The following Figure 5.4 provides a checklist of some of the associative relations that have been inserted in the thesaurus and integrated thanks to the application of the word-embeddings and pattern-based configurations, for the sake of clarity the implementation of new associations are given in blue colored text in the right-hand-side of the table.

5.1.1.4 Scope notes

Figure 5.5 offers some examples of the notes that supplement the specification of the Cybersecurity terms included in the Italian thesaurus giving authoritative definitions. Most of these latter come from the standards NIST 7298 r2 *Glossary of*

term	associative relation
Anti-Spoof	Spoofing, Anti spam, Cyber defence, Trojan horse, Koobface, Cyber risk
Brute Force Attack	Password
Biometrics	Biometrics vulnerabilities
Blockchain	Security properties, Cryptocurrency
Botmaster	Botnet, DDoS
Cyber insurance	Cyber terrorism
Trojan horse	Crypto miner malware,Anti spam, Cyber defence, Hacking, Risk management
Crackers	Hacking, Hackers, Backdoor
Criptography	Decryption
Cyber bulls	Cyber bullism, Cyber trolling, Dark web
Cyber defence	Cyber security, Cyber forensics
Cyber war	Dark web, Cyber terrorism, Cyber warriors, Cyber fighters, Cybersecurity, Cyberweapon, Cyber terrorists
Cyber laundering	Dark web, Cybersecurity
Cyber threats	Hacking, Cyber risk assessment, Dark web, Cyber Threats
Cyber terrorism	Intelligence, Privacy, Integrity, Availability
VBS	Cyber weapon
Cyber Threat Intelligence	Melissa
Cyber-hygiene	Cybersecurity
Cybercriminality	Cyber attacks, Hacking, Cybersecurity, Dark web, Cyber forensics
Cybersecurity	Cybercriminality, Cyber defence, Cyber terrorism, Cyber-hygiene, Cyber espionage, Hacking, Privacy, Authentication processes, Cyber war, Cyber harassment, Cyber bullism, Cyber laundering, Cyber theft, Cyber Threat Intelligence, Cyberbullism, Hackers, Blockchain
Zero-day	Exploit Zero-day, Trojan horse, Anti spam, Hacking, Privacy
Hackers	Hacking, Crackers, Cybersecurity, Cybercriminality
Phishing	Spoofing
Malicious software	Antispyware software, Firewall, Exploit, Keylogger
Spam	Spoofing, Virus
Virus	Spam, Antivirus software
Vulnerabilies	Cyber attacks, Cyber risk assessment, Exploit, VAPT
Worm	Sobig

Figure 5.4: Related terms

Key Information Security Terms, ISO 27000:2016, White Book of Cybersecurity and the Glossary of Intelligence.

5.2 Ontology conversion

A large number of studies on the possibility of re-engineering a thesaurus into an ontology have proved that migrating the terms from a thesaurus to an ontology structure aims at offering a semantic arrangement which is more readable by the informative systems thanks to the exploitation of Resource Description Framework (RDF), Resource Description Framework Schema (RDFs), the Web Ontology Language (OWL) languages, which commonly provide a more flexible and interoperable way of sharing information. The conversion of the thesaurus terms structure into a higher form of conceptualization regarding a representation of a knowledge-domain is based on the purpose of providing a more elaborated framework for the system of semantic connections to be explained among domain-specific concepts. Indeed, OWL language allows to apply to domain-oriented concepts specific semantic properties, e.g., (i) the disjunction occurring in two sets in a way that concepts can be taken into account separately; (ii) the functional association of a property to a class, to which follows the attribution to unique identifiers; (iii) the transitive property among classes that can be considered as an aide for more detailed and descriptive semantic outlines. The main feature of OWL consists in its higher formalism mechanism in delineating the framework relating to technical fields of knowledge and its inner automatic reasoning systems which can support the inference processes implementation on the described knowledge concepts. Many works on the re-engineering process from a thesaurus to an ontology have demonstrated that the terms contained in the first semantic resource can be migrated by adapting them and their relations in the properties proper to ontologies, as Lauser (2006) for the AgroVoc domain, or Cardillo (2014), Nowroozi [118], and Kless (2012) highlight in their works. The reasons why the ontology is considered a more specific and explicit form of knowledge representation is given by the fact that it provides a deeper and more explicit form of description of the connections existing among concepts belonging to specialized domains of study. In particular, the contrasting analogy with the thesaurus relies on the precise way the ontologies adopt in representing the associative relations by assigning to the relations elapsing among classes a specific *Object Property* with a specific URI able to provide univocality in terms of terminology connections precision.

Indeed, this issue is addressed in a large number of studies, such as the one carried out by Van Assem *et. al.* (2004) with their experiment consisting in the conversion of the MeSH and WordNet thesaurus to RDF and OWL language. The authors underline the constraints a thesaurus exhibits in the way it manages the RT connections among terms. The migration process they take into account is

term	note
Back Door	Typically unauthorized hidden software or hardware mechanism used to circumvent security controls.
Biometrics	A measurable physical characteristic or personal behavioral trait used to recognize the identity, or verify the claimed identity, of an applicant. Facial images, fingerprints, and iris scan samples are all examples of biometrics.
Botnet	Network composed of devices infected by specialised malware (malicious bot) and controlled by a so-called botmaster
Trojan Horse	A computer program that appears to have a useful function, but also has a hidden and potentially malicious function that evades security mechanisms, sometimes by exploiting legitimate authorizations of a system entity that invokes the program.
Cyber defence	A set of doctrines, organizations, activities aimed at preventing, detecting, limiting, and fighting the effects resulted by the attacks in and within the cyber-space, affecting one or more of its components.
Cyber threats	Any circumstance or event with the potential to adversely impact organizational operations (including mission, functions, image, or reputation), organizational assets, individuals, other organizations, or the Nation through an information system via unauthorized access, destruction, disclosure,modification of information, and/or denial of service.
Cyber space	A global domain within the information environment consisting of the interdependent network of information systems infrastructures including the Internet, telecommunications networks, computer systems, and embedded processors and controllers.
Cybersecurity	The ability to protect or defend the use of cyberspace from cyber attacks.
Dark web	The parts of the web containing data not publicly indexed or whose access is protected through anonymous networks. It is believed that about 80% of the contents present today on the world wide web are hidden in these parts.
Exploit	A program that allows attackers to automatically break into a system.
Hackers	Unauthorized user who attempts to or gains access to an information system.
Cyber attacks	An attack, via cyberspace, targeting an enterprise's use of cyberspace for the purpose of disrupting, disabling, destroying, or maliciously controlling a computing environment/infrastructure; or destroying the integrity of the data or stealing controlled information.
Phishing	Tricking individuals into disclosing sensitive personal information through deceptive computer-based means.
Ransomware	Malware that restricts the use of a device, for example by encrypting data or denying access to the device itself.
Social engineering	A general term for attackers trying to trick people into revealing sensitive information or performing certain actions, such as downloading and executing files that appear to be benign but are actually malicious.
Malware	A virus, worm, Trojan horse, or other code-based malicious entity that successfully infects a host.
Spam	Electronic junk mail or the abuse of electronic messaging systems to indiscriminately send unsolicited bulk messages.
Vulnerabilities	A weakness in a system, application, or network that is subject to exploitation or misuse.

Figure 5.5: Scope Notes for terms in the thesaurus

characterized by four steps, (i) preparation, (ii) syntactic conversion, (iii) semantic conversion and (iv) standardization. The first step is based on the analysis of the thesaurus considering the:

> "– Conceptual model (the model behind the thesaurus is used as background;
> knowledge in creating a sanctioned conversion);
> – Relation between conceptual and digital model;
> – Relation to standards (aids in understanding the conceptual and digital model);
> – Identification of multilinguality issues." (2004:3)

The conversions considered by the authors are of two types:

1. **Syntactic**: involves a structure translation of all the semantically relevant units contained in the thesaurus into RDF syntax. The preservation of the source structure is guaranteed, among others, by using RDF(s) constructs for the definition of classes and subproperties **rdfs:label**; xml for datatyping; maintenance of the original forms of naming entities, with specifications given by **rdfs:comment** or **rdfs:seeAlso**; avoiding redundancy.

2. **Semantic**: in this conversion step the authors evaluate the explications of properties through RDFS and OWL constraints: for example, the thesaurus semantic relation referring to the Broader Term (BT) is given by the **owl:TransitiveProperty**, while the Related Term (RT) into **owl:SymmetricProperty**. The more fine-grained approaches address to consider the thesaurus as a *metamodel* converting, for instance, an instance in a class **rdfs:class** and the BT hierarchy as a class hierarchy transposing it in RDFS syntax as **rdfs:subPropertyOf, rdfs:subClassOf**[3].

The standardization step implies a mapped comparison with a standard schema. Similarly, for this work project the syntactic and semantic conversion have been carried out with the purposes of maintaining the schema model proper to the Italian thesaurus meant to be migrated. In conformity to this objective, the terms contained in the source metamodel thesaurus for the Cybersecurity domain have been converted into classes and their relations in *Object* and *Data Properties*.

Despite thesauri's actual reliability in structuring a deep domain-specific configuration and developing a dense network of semantic relationships, they prove to be quite less precise in portraying the exact type of connection linking terms that can formalize the conceptual domain-oriented framework. It has to be noted that even transposing the terms tissue within the controlled vocabulary to classes, entities and properties schematization conceptual ontological modelling, as in

---

[3] in this work the principle of taxonomical information structure has been followed by transposing the hierachical thesaurus' configuration in **is-a** and **kindOf** relations

the case of a thesaurus' reliability check, the validation carried out by domain experts has to be kept in mind to verify the authoritativeness range the resources should hold to be shared within the community of people working in the field of knowledge under analysis. In the sections below a description of the structure given to the ontological conformation is provided, as well as examples of the ways ontologies can prove to offer higher specific connections among the taxonomic chains with respect to the associative relations which are present in the thesaurus. Also in the ontology, the source language with which the research project started has been the Italian one, and successively, as was the case for the thesaurus, the alignment with the English language has been carried out with the use of the retrieved equivalents.

5.2.1 Structure

The structure of the Italian Cybersecurity ontology, developed with the Protégé platform[4], to be included in the services of the OCS web portal for Cybersecurity reference guidance, has been created under the basis of the Italian thesaurus metamodel structure for the same domain of study and reflects its inner connection outline among terms transposing them in a higher conceptualized conformation able to guarantee a more fine-grained explicitation of the knowledge-domain concepts. The following Figure 5.6 shows the main structure of the ontology derived from the thesaurus' second conformation that implemented and adjusted the associations given by the patterns and word embedding models, such as the modification executed for the *cyber attacks* concept that has become a **sub-class of** *cyber crime* and not of *cybersecurity* as it is in the first version of the thesaurus which was agreed upon with the experts.

As previously mentioned, the main objective to be achieved in migrating the terminology thesaurus framework in an ontological taxonomic structure is to facilitate the highly articulated description of the semantic associations occurring among classes of concepts. Details of the ontological configuration developed by using Protégé are provided in Table 5.1. The perspective in using OWL and RDF(S) languages relies on the possibility of exploiting the query system tasks included in the ontology software applications, through the plugins that allow SPARQL operations mapping with external drivers. SPARQL queries system enables to activate reasoning engine processes to infer semantic connections from several resources given in input as conceptual models in order to automatically enhance the terminology strength with authoritative information coming from trustworthy sources.

To sum up, one of the main advantages given by the re-engineering process to an ontology infrastructure, considered a valid transposing method to widen the semantic scope precision of the semantic connection system within the Italian

---

[4]https://protege.stanford.edu/

Figure 5.6: Ontology graph for the Italian Cybersecurity thesaurus conversion

thesaurus for the Cybersecurity domain, is the highly structured framework it offers to clarify the associative relationships between the terms of the thesaurus [88].

The two semantic systems, the thesaurus and the ontology, offer a clear distinction in the way they manage and represent specialized knowledge-domains, such as Cybersecurity, and the main difference can be found in the structure of the semantic associative relations that are more explicit in the ontology.

a. Object Properties

The *Object Properties* expressed in a OWL built-in class as **owl:ObjectProperty** are applied to represent a binary correlation among entities and describe classes by the aide of restrictions. Figures 5.7 and 5.8 depict the relationships linked to the idea of opposition, i.e., *spoof* and *antispoof*. In particular, the opposition matching in the thesaurus' knowledge management is provided only through the associative relation (RT) between the two terms, and in order to better define the term's semantic extent a definition is given (within the black square), i.e., Scope Note (*SN*), that provides a official definition retrieved from the authoritative source textual documentation or from the gold standards.

Differently, the two terms representing the relationship of opposition are represented in the ontology using the **rdf:type owl:ObjectProperty**, subclass

Table 5.1: Cybersecurity ontology metrics on Protégé platform

Metric	Total
Axiom	644
Logical axiom count	332
Declaration axioms count	229
Class count	155
Object property count	35
Data property count	7
Individual count	31
Annotation Property count	5
CLASS AXIOMS	
SubClassOf	157
DisjointClasses	23
OBJECT PROPERTY AXIOMS	
SubObjectPropertyOf	33
ObjectPropertyDomain	38
ObjectPropertyRange	36
DATA PROPERTY AXIOMS	
SubDataPropertyOf	6
DataPropertyDomain	8
INDIVIDUAL AND ANNOTATION AXIOMS	
ClassAssertion	31
AnnotationAssertion	81

of the RDF class **rdf:Property**, "HasAsContrary," which supports the explicit form of union these concepts share, considering the *Domain* and the *Range* as linked by a precise customized relationship that has been determined by applying the pattern-based recognition processes and word embedding models throughout the source corpus documents. Another example of the transposition carried out by the associations which have been structured in the Italian Cybersecurity source thesaurus, which now works as metamodel to the construction of an improved way of representing specialized terminology, is shown in Figure 5.12, which highlights the connections the concept *backdoor* exhibits with *malware* and *crackers* with the use of complementary patterns configurations on Protégé console; while Figure 5.9 shows the connection *exploit* occurring between the concept *vulnerability* and *cyber attacks* being the first one the *Range* and the second the *Domain*, and the relation *monitor* between *cybersecurity* and *vulnerability*, which still is conceived as *Range*. Another type of connection is given by

Figure 5.11 representing the *detection* the *antispyware software* run over the *malicious software*. Finally, Figure 5.10 presents the structure the ontology allows in representing the *actors* of a specific event: *hacking* has as its actors *hackers* and *crackers*. All the above relations are given in the thesaurus with the associative relations, and this can demonstrate the flat visualization it exhibits in contrast with the ontological representation of the interrelating connections.

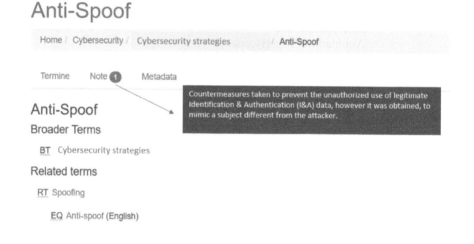

Figure 5.7: Thesaurus representation of the semantic relationship that describes opposition

b. Data Properties

The re-engineering process has been carried out also to organize in a more precise way the links among concepts, as the ones referring to the **attributes**. The OWL built-in class to define this type of relation is the **rdf:type owl:DatatypeProperty**, subclass of the RDF class **rdf:Property** and the connection it allows is between individuals and literal data, such as, strings. Figure 5.13 shows the dissimilar ways of visualization for the attributes associated to the concept *security properties* in the ontology semantic tool, i.e., *integrity, authenticity, confidentiality, availability, reliability, non-repudiability, and privacy*, while in the thesaurus they appear to be related to the hyperonym BT "Data" and are represented as its specific terms, i.e., the NT (Iso 25964-2:2013), as Figure 5.14 shows below.

c. Individuals

This type of ontology entities in the thesaurus are covered by the instantiative relationship NTI; in the ontology, on the contrary, these representative cases are

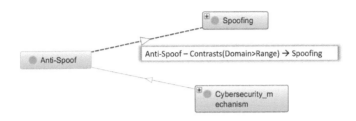

Figure 5.8: Protégé representation of the semantic relationship that describes opposition

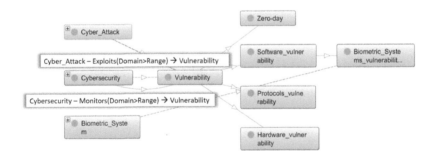

Figure 5.9: Ontology representation for vulnerability connections

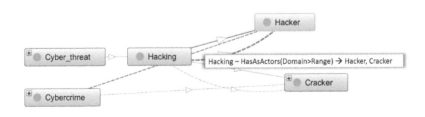

Figure 5.10: Ontology representation for actors: Hacking with hackers and crackers

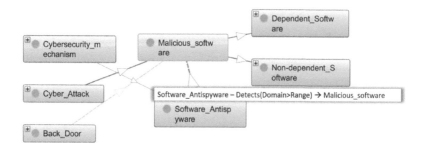

Figure 5.11: Ontology representation of antispyware and malware connection

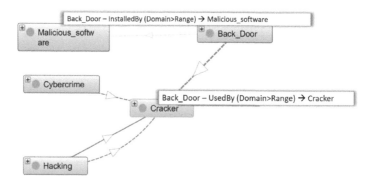

Figure 5.12: Additional ontology Object Properties through patterns path-variables

Figure 5.13: Ontology representation of Security properties as *Data Properties*

Security Properties

Broader Terms

BT   Data

TT   Cybersecurity

More specific terms

NT3   Authenticity
NT3   Confidentiality
NT3   Availability
NT3   Integrity
NT3   Non-repudiability
NT3   Privacy

Related terms

RT   Blockchain
RT   Cybersecurity mechanisms
RT   Spoofing

Figure 5.14: Thesaurus representation of Security properties as hierarchical relations

solved by assigning to a class concept a range of *individuals* that represent the instances, specific names, of a domain concepts in a more precise way with respect to the thesaurus configuration. The following Figures 5.15 and 5.16 illustrate the attribution of the individuals in the case of *crackers* being specified as *script kiddies* and *lamers*, and the names belonging to some of the main *independent malicious software*, such as *zombie, worm and ransomware*.

As mentioned before, the ontology has been forged under the basis of the thesaurus structure to organize the Italian terminology about Cybersecurity. For this reason, given the primary scope of the ontology the higher terminological explicit systematization that leads to a more appropriate semantic representation of the domain concepts, the OWL sublanguage selected is OWL Lite.

To enhance the accuracy in the representation of the Cybersecurity domain, the ontology has also been integrated with the new semantic elements resulting from the implementation of path-variables sets (Roesiger, 2016). The selection process already executed for the purposes of the thesaurus semantic relationships' automatic detection has equally been employed for the ontology's semantic expansion as Figure 5.17 demonstrates below, collecting causative verbs and hyperonymy as well as synonymy structures, with a specific attention to the morphological and syntactic constructs proper to the domain of Cybersecurity in order to refine the associative connections among the concepts (Condamines, 2006) as much as possible.

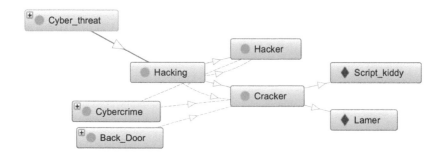

Figure 5.15: Crackers' individuals

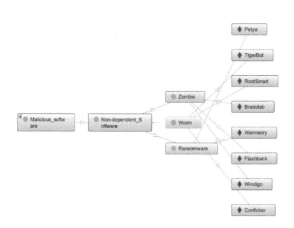

Figure 5.16: Individuals for malicious software representation

Associative relationships list			
Aggirare (bypass)	**Attaccare (attack)**	**Sfruttare (exploit)**	**Attivare (activate)**
worm > software antivirus	exploit sql injection> web applications	crackers > vulnerabilities	backdoor > malware
cracker> cybersecurity	script kiddies > DDoS	trojan horses > vulnerabilities	trojan > cyber attacks
virus > cybersecurity	network file systems > DNS spoofing	spam > botnet	payload > virus/worm

Figure 5.17: Associations retrieval in ontology by patterns configurations

5.3 Discussion

The specific use case, for which both thesaurus and ontology for the domain of Cybersecurity in Italian language have been realized, has piloted the realization of these semantic monitoring and management resources. Specifically, the underlying objective of this study outlined in the present book has been that of guaranteeing to common users, experts of the domain, institutions and official political and administrative organisms, a controlled way of understanding a highly specialized terminology in a structured and interoperable way while ensuring that the semantic tools realized for this purpose are meant to keep being as reliable as possible with respect of the quantitative and qualitative level of information regarding the domain. The achievement of a trustworthiness status relied on the different ways the thesaurus, and successively the ontology, offers a correspondence to the knowledge-domain through semantic relationship networks. This connecting framework aims at interlinking terms with each other in order to enable users to have a detailed and supported guidance to the specific degree of terminology characterizing this specialized area of study. Both for thesaurus and ontology the system of relations occurring among terms and concepts comes from the implementation of algorithms executed on the authoritative Italian source corpus compiled for the Cybersecurity domain and from the gold standards guidelines mapping task; together they create a baseline for the arrangement of hierarchical, equivalence and associative interrelations. One of the main disadvantages which has been frequently highlighted in the previous sections is the flat visualization thesauri provide in associating specific terms to each other, an issue that can lead to some vagueness in the matching processes.

Therefore, the creation of another semantic resource, i.e., the ontology, compensated to the lack of explicit semantic descriptions, as well as terminological relations between the domain concepts thanks to the possibility of inserting properties that can be subjectively customized by making the conceptual interrelation task more accurate.

It has to be noted that even though the two semantic resources are part of the same knowledge organization system group and share some key functions in the management tasks for a specific area of study, such as the fact that they both support the information retrieval processes, their scope differs: the ontology is characterized by an higher formal knowledge representation with respect to a domain by the application of the functionalities proper to the RDF and OWL interoperable languages it exploits, e.g., explicit relationships between concepts, disjunctions, data properties for each concept or instance and by restrictions, that limit the possibilities of detecting ambiguity in the conceptual semantic representation and in the applicability within a peculiar use case scenario. Differently from the more extensible way of adjusting the conceptual model in the ontology, given by its scalability and reusability, as pointed out by Blumauer and Pellegrini [8], which leads to superior semantic richness thanks to an internal deeper formal description of concepts, the thesaurus is characterized by a structural strict way of formalizing specialized terminologies. However, compared to the ontology, a thesaurus is a semantic tool that avoids ambiguity with the aid of the synonym control [132], which supports the decision-making process related to the preferred terms detection meant to represent a domain-specific concept. This can guarantee a higher form of accuracy in the selection of the best candidate preferred terms, to then be migrated in the classes of the ontology. Moreover, the rigidity that distinguishes the thesaurus is a feature that can support the standardization of technical information, and, thus, a process of unification and sharing the terminology of a specific field of knowledge. This aspect can help through many concrete usages under the basis of an homogeneous understandable and semantically connected language. Nonetheless, as largely proven in the literature, thesauri can become a valid source metamodel that can be used to populate the ontology by transposing the fixed standardized structure determined for the domain-specific terms in RDF and OWL language constructs, which guarantee a better form of interoperability among the informative systems, and, by consequence, an open accessible form for a structured specific knowledge representation. Despite this, one of the strengths of the thesaurus compared to the ontology, when used in a specialized domain, is its greater capacity to eliminate ambiguity between the terms through the use of synonymy control (Zeng, :::) and the choice of preferred terms, compared to non-preferred terms for representing the concepts. This guarantees a standardization of technical terms in specialized domains, which can help in the process of unifying, and, by consequence, sharing, a specific field of knowledge's terminology.

Chapter 6

Semantic enhancement and new perspectives

In this chapter some of the techniques already tested out to improve the semantic richness of the Cybersecurity developed semantic tools will be presented, as well as others that represent solid perspectives aiming at automatizing the semantic enhancement within the term structures. For what regards the objective of enriching the semantic tissue proper to the terminology management resources, the preliminary methods that have been experimented have addressed to the realization of multilingual semantic tools in order to guarantee a broader coverage of the terminology of this specialized field of knowledge. As mentioned in Chapter 4, the effort implied for the analysis of the parallel textual documentation would have been considerably elevated, hence, the translation represented a key solution to embrace the objectives based on the realization of bilingual semantic monitoring tools. The second approach relies on the application of other techniques targeted to discover new information starting from authoritative sources as to guarantee a constant updated version of the terminology conceived as representative for the domain of study.

6.1 Multilingual alignment

One of the last phases carried out for the purposes of this joint research project with the IIT-CNR institution, has dealt with the alignment of the terms contained in the thesaurus with the English ones to be successively imported in the ontological framework. The procedure has addressed the retrieval of the equivalent semantic forms to be integrated in the thesaurus and in the ontology in order to

DOI: 10.1201/9781003281450-6

guarantee a multilingual authoritative toolkit for the understanding of the Cybersecurity technical domain of study in Italian and English languages. As aforementioned in Chapter 4, the corpus in the English language represented an empowered documentation for the state-of-the-art discovery, coverage extent level analysis and for checking the translation correctness in the equivalents detection system. The tasks covered to collect English equivalent terms have mainly been two:

1. Manual translation operations: this type of equivalences transposition has been addressed to the translation into Italian of the terms selected after the mapping matching tasks described in Chapter 4. Similarly to what has been done in the English-Italian translation process, which has been carried out by using IATE platfrom, a valid system for translators and terminologists which offers the possibility to consult the main applicative specific contexts in which terms are effectively used, e.g., *virus* in Medicine area or *virus* in ICT security field, the Italian-English translation process of the filtered out candidate terms to be included in the thesaurus, and then transposed as classes, individuals and OWL properties in an ontology system, started manually using the same web platform to guide the adherence of the proposed translation to the specific use case, i.e., the ICT security area.

2. Software-aided automatization: the second methodology employed for the purposes of retrieving English equivalents, and thus widening the coverage threshold of the semantic tools meant to manage and represent the knowledge-domain of Cybersecurity, addressed to the use of computer assisted tasks through the alignment software Alinea [103] which is a concordance bilingual automatic sentence aligner optimized for several languages, that allows a manual modification, works with complex regular expressions and XML tags constructs, extracts bilingual lexicon and executes evaluations with precision and Recall. This software covered the step of equivalents detection and supported the validation processes for the terms which had undergone a manual transposition from English to Italian.

In both cases, manual and automatic bilingual correspondence retrieval, there is a substantial issue to take into account, that is the mixed language nature permeating the domain of Cybersecurity, characterized by the following features:

■ multisciplinarity: the domain of Cybersecurity involves ICT and sub-ares (Audiovisual techniques; Computer software; Electronics; etc)

■ specificity: the terminology proper to the Cybersecurtiy field of knowledge is full of technical and standardized terms;

■ cross-field coverage: the domain applications fall into the computer science field, legislative systems, regulations, social media world.

This leads to the fact that, sometimes, some of the terms included in this field of knowledge do not find a precise reference translation in the target language, or they cannot even be translated. The reasons why untranslatability in terms does occur are essentially three:

a. the terms become themselves accepted because of their common frequent usages within the targeted contexts in their foreign language form. This is the case for example of *cyber-hygiene*, or *virus*, *password*, *phishing* and *malware*;

b. it is not possible to define a translation for certain terms, such as *honeypot*, *smishing*;

c. if the pure meaning is encapsulated and well-integrated in the original form and translating the terms would imply a conversion or an upheaval of significance, as for example in the case of *social engineering*, *blockchain* and *bitcoin*; in this case there is the risk of undermining the meaning if the translation implied the splitting of complex lexical units, loosing the radical sense of the terms.

6.2   Monitoring of terminological representativeness RDF Blank nodes reasoning

The following Figure 6.1 provides a hypothetical model-based graphic representation of a possible updating system derived from the sources consulted for the creation of the reference corpus through a RDF graph. RDF language, allows to describe through subject-predicate-object triples, incomplete variables according to which some properties of a certain state of things are known, and without being aware of what these triples concretely represent, it is possible to predicate some of those properties. An RDF statement is a description of a resource and it is constituted by a subject, a predicate and an object that identifies the value of the property. This system of incomplete reasoning, through the use of the so-called *blank nodes*, proves to be useful in describing the way by which the reports of the updates from the authoritative source can be managed before starting a reconsideration of the terminological structure of the thesaurus and of its relation network. In detail, it is possible to consider an possible document (DOC X) created by a hypothetical source (SOURCE X), that can be the subject (rdf:subject) of the change notified by an alert through a report type (rdf:type) of a statement (rdf:statement) that has as its predicate (rdf:predicate) the description of a state of things having as object (rdf:object) a text. The alert should have to be received starting from the establishment of determined criteria for what concerns the provenance selection of new information.

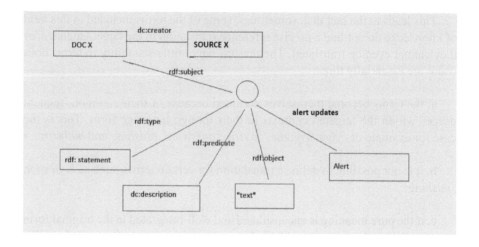

Figure 6.1: RDF node

A viable way to retrieve new semantic information in order to expand the semantic resources for the Cybersecurity domain coverage extent is to consider variants extraction within the terminology processing lists. As already shown in section 4.1.2.2, the variants detection tested with the aide of the semi-automatic term analyzer software has represented an efficient methodology towards the enhancement of the term structure and the improvement of the specificity referring to technical lexicons under study, as well as a technique to automatize the hierarchical and synonymous levels in the thesaurus. The path that can be followed in the perspective of automatically empowering the semantic outline of the thesaurus and the ontology is, hence, related to the discovery of new terminological variations. This leads to the possibility of accessing novel sets of variants lists given by updated corpora related to the corresponding knowledge-domains. Starting from the assumption that new information can proceed either from a deeper lexical analysis of the compiled source corpora or, more specifically, from the integration of new documents published by the same a priori selected authoritative reference sources, the variants recognition system would then be capable of highlighting new terms retrieved in the updated documents of the source corpora and their variants detected by the semantic software, creating the basis of an enhanced framework of new specialized information detection. In this way, not only the terms extent could be empowered and, eventually, expanded, but also the semantic relations network could be intensified providing a higher level of accuracy in the information contained in the semantic tools and in their representation of knowledge-domains to common users and experts.

SPARQL query expansion

It is worth mentioning the study conducted by Shamsfard (2003) on the automatic ontology construction based on the Persian language proving that an ontology could be considered as a semantic tool able to represent in a more explicit way the network of semantic relationships occurring among concepts by the means of unique forms, i.e., Universal Resource Identifier (URI) names, and specific customized properties conceived to depict those connections. The definition provided by the authors related to the ontologies summarizes its conceptual structure:

> "The ontology is defined by O=(C, R, A, Top) in which C is the set of all concepts (including relation concepts), R is the set of all assertions in which two or more concepts are related to each other, A is the set of axioms and Top is the most top level in the hierarchy. R itself is partitioned to two subsets, H and N. H is the set of all assertions in which the relation is a taxonomic relation and N is the set of all assertions in which the relation is a non-taxonomic relation."

For the purposes of enlarging the qualitative and quantitative level of the conceptual structure contained in the Cybersecurity ontology, SPARQL queries can be exploited. The idea behind the use of SPARQL query system is to start compiling a indoor driver of new information resulted from a secondary analysis of new documentations inside the source corpora, and then apply a mapping between the pre-developed ontology and the structured table in order to automatically populate the former with new classes inferred by the reasoning operators. Indeed, the plugin OnTop[1] works in this direction, mapping the concepts belonging to a class to the data frames which have been compiled on the external drivers to be connected. In this way, the matching operations granted by the SPARQL queries syntax facilitate the incorporation of new data imported from the updating process implemented over the source corpora. As a result, the ontological system could be automatically populated and be more representative since it contains new specific concepts of the domain.

Social media semantic detection

One method that can be efficient for empowering the semantic updated systematization for the domain of Cybersecurity, and improve the representativeness threshold of its semantic monitoring resources, can be that of relying on the tweets published by authoritative sources in order to isolate meaningful lexical elements representing triggers of new information. Tweets can provide an additional semantic information to be imported within the semantic system

---

[1] https://ontop-vkg.org/

structures with respect to the compiled source corpus, because their data streaming is updated each time with news about the cyber context episodes with a delimited fixed length posts of 140 characters, that can help in setting out a precise parsed morpho-syntactic pattern scheme to retrieve new specific information to study. The sources to be taken into account should have to comply with the criteria outlined for the composition of the Italian Cybersecurity source corpus, i.e., authoritativeness and reliability features. Therefore, a list of trustworthy profiles needs to be arranged going from, for example, NIST and ENISA accounts for the English language, or CINI Consortium, Clusit, ITASEC for Italian. Classifying the lists of authoritative cyber profiles, processing tweet strings and elaborating specific distribution similarity approaches to detect new information about the terminology of the domain, can be performative in augmenting the coverage threshold for both semantic tools. Twitter, among the other social media, can offer some advantages among which:

1. analysis of the crowds' language, a non-structured way of natural language arrangement; e.g., colloquialisms, jargons that are used in the social media world, hence another typology of specific terminology to be studied [94];

2. the major fluidity of language, the rapidity traced in the new updates, which can be confusing at first, but can bring benefits in observing and retrieving continuous updates to take into consideration to apply modifications on pre-elaborated semantic resources [89];

3. the fixity of the tweets' length and the rules that can be applied in the crawlers filters specifications, such as, the source, the keywords, the hashtags, the geographic areas, the number of retweets, the language, the accounts mentioned and to whom a tweet is addressed for example [68];

4. the possibility to create lists that can help in sorting out posts of interest and create an upstream categorization process to extract the desired information.

Given these premises, Twitter can represent a valid support tool to launch queries to collect specific outputs [101]. The text mining process implemented over tweets involves NLP techniques for discovering semantic units that can be analysed towards the empowerment of the thesaurus and the ontology for Cybersecurity [15]. Indeed, the purpose of this prospect activity is to use the information, as tweets' contents, originated by selected authoritative profiles and apply NLP processing tasks to extract meaningful linguistic constructs that can help in the discovery of new semantic information, such as new names assigned to malware by specialized social media authorities, and ameliorate the terminology representativeness.

The crawling that can be started using Twitter text mining procedures can configure the strings to be computationally processed as the following:

DATE 06/11/2020 **TIME** 04:38 **SOURCE ENISA TWEET** During the pandemic, phishing scams increased unprecedentedly. Don't click on links or download attachments unless you are confident about the source of an e-mail. To learn more, check out the latest ENISA Phishing threat report **HASHTAG#** *#ENISATL20.*

This tweet highlights that compounding **phishing scams** could be a new type of information to select in the perspective of including it as new candidate terms meant to be validated by a group of domain experts.

DATE 21/10/2020 **TIME** 02:10 **SOURCE** ENISA **TWEET HASHTAGS#** *#Malware* is standing strong as Cyber Threat in the EU followed by Web-based attacks & #Phishing. Find out more about the top cyber threats during a period when *#COVID19* fuels attacks on homes, businesses, and critical infrastructure. *#EnisaTL20.*

In this case **Malware – Cyber Threat – Web-based attacks – Phishing** appear to be newly correlated and this can represent a new type of semantic information to take into account to enhance the thesaurus and ontology frameworks.

DATE 08/10/2020 **TIME** 11:06 **SOURCE** @NISTcyber Cybersecurity @ NIST A tip we shared in our guide for protecting connections, includes using a combination of security software, such as antivirus software, personal firewalls, spam/ web content filtering, & popup blocking to stop most attacks **HASH-TAGS#** *#BeCyberSmart* **MENTIONING** via @IDTheftCenter @StaySafeOnline RT @StaySafeOnline: Q1: One of the key messages for this year is: If You Connect It, Protect It. Our personal lives are getting more connected (smart homes, wearables, etc.). What are some ways we can protect our connections at home? **HASHTAGS#** *#BeCyberSmart* **MENTIONING@** IDTheftCenter.

Finally, here the pattern-based element *such as* is notifying that among the security software the terminology resources should contain *antivirus software, personal firewalls, spam/web content filtering, & popup blocking* as specific terms, adding the extra information preceded by the verb **stop** with the term *attacks*.

As it can be observed in the previous examples, the posts published by ENISA and NIST, two outstanding authorities in the field of Cybesecurity, provide many tracks to cross towards a semi-automatic specification of the knowledge-domain terminology under study. The achievement of realizing semantic resources that can be representative for the domain of interest is in this manner improved by aligning the social media mining semantic information contained in the tweets, which follows specific rules, to all the procedures carried out for the realization of the thesaurus and the ontology, and thus being in line with twofold languages:

- **structured languages:** represented by legislative and scientific domain-oriented magazines;
- **unstructured language:** represented by the social media lexicon which in the case of Twitter observes specific syntactic rules in writing the posts.

In both cases the NLP techniques are essential to perform automatic sophisticated tasks over big data sets and start the new semantic units recognition process to empower the coverage measurement guaranteeing the quantitative and qualitative representativeness of specialized fields of study by semantic resources such as thesauri and ontologies.

Chapter 7

Conclusion

In this research the development of two semantic monitoring tools, a thesaurus and an ontology, for the Italian Cybersecurity specialized field of knowledge management and representation is described, together with a presentation of the main NLP techniques exploited to discover which can be the best condition for the two KOS systems to be representative with respect to the information domain under study, particularly referring to term variation. This book presented an exploratory research towards the delineation of a methodology – meant to be used as a model for other specific domains of interest – able to compute a minimum quantitative and qualitative coverage threshold for a thesaurus in order to become, and keep being, a reliable representative means of terminology control for a highly technical domain such as that of Cybersecurity.

The research activities have been carried out in collaboration with the IIT-CNR, and in particular with the CyberLab, the leading Italian group that has realized an online accessible platform, called Cybersecurity Observatory (OCS), to guide the users in the comprehension of the main and latest phenomena occurring in the field of Cybesercurity. The thesaurus and the ontology, objects of this work, have been implemented with a view of integrating them among the OCS services, namely in the semantic sections as guidance for the highly specialized lexicon of Cybersecurity, full of English loans and constantly subjected to changes over the time. The application domain of Cybersecurity is, for instance, characterized by a multidisciplinary complex convergence of sub-areas, such as that of Information and Communication Technologies (ICT) with its sub-partions, as the audiovisual techniques, computer software, electronics ones. Being a domain that appears to present an highly specialized lexicon, the book proposal is to manage its terminology in order to facilitate the understanding process by a community of users, and to standardize an homogeneous technical application of the domain terms'

DOI: 10.1201/9781003281450-7

meanings in several contexts of usage. The peculiarity of the domain of Cybersecurity lies in its cross-fielding thematic coverage. In fact, for this project different typologies of textual documentations have been consulted to build a sound corpus to represent the knowledge-domain spacing from the computer science sphere, the legislative system, the regulatory texts to a more popular section of scientific and sector-oriented magazines issues.

Processing the cross-fielding terminology proper to the domain of Cybersecurity has been supported by creating formal semantic management systems, a thesaurus and an ontology, to provide a structured configuration of the technical lexicon within this area of study. The thesaurus, subsequently re-engineered into an ontology system, has been chosen as the semantic monitoring tool rather than other possible knowledge management systems, such as, taxonomies, glossaries, or classification systems. Despite the presence in the literature of ontological resources on the area of study belonging to the Cybersecurity, mainly in the English language, or glossaries published by trustworthy Italian and English sources, and also non-controlled lists of words providing definitions to some of the domain-oriented concepts, the thesaurus has been preferred for its semantic rigidity that confers a high standardization when structuring the domain specific terminology to be managed that can facilitate a process of reusability for disparate needs, and for the possibility to systematize specialized terminologies within a interconnecting network of hierarchical, equivalence and associative relations among representative domain terms. Indeed, the thesaurus allows the management of a specific knowledge field because of its systematic order that enables the creation of terminological connections between the terms, meaning the creation of an extensive network of semantic relations as to choose the best proper terms to guide the comprehension of the knowledge-domain. This semantic tool, together with its transposition to an ontology taxonomy, aimed at guaranteeing the check of terminological coverage for the domain of Cybersecurity as well as organizing a base of awareness in this area and a support for organizations for what concerns the operations of information access, retrieval and processing. Being the OCS web platform a portal that has been developed by an Italian institution, one of the preliminary requirements the realization of these semantic means of terminology control has been focused on was to create an **Italian** thesaurus for the Cybersecurity domain. The main innovative feature brought forward by this study is that this type of semantic tool, i.e., a thesaurus, referred to this domain able to gather as many concept and term as possible that are needed to classify documents and/or resources of the area with the use of semantic relations did not officially exist in the Italian language. For this reason, the terminology has been firstly structured starting from the Italian language framework, and only after having set up a semantic configuration validated by the experts of the domain in Italian, the incorporation of bilingual terminological resources could have been started. Therefore, the language from which the project of this work took its ground is the Italian one, with which the processing of the corpus texts, mapping procedures, NLP techniques related to the semantic relations discovering, have been lately adapted to the subsequent English equivalence detection process. The perspective of taking into account English sources has as its end the conceptual analysis the meanings terms assume in different geographic contexts and the possibility of matching or not the semantic boundaries from a source language to a target one. Specifically concerning this latter issue, the fact that the domain of Cybersecurity is characterized by a high number of foreign English terms is an important element to take into account when incorporating some of the terms in a structured interoperable resource, which is supposed to observe the linguistic agreement of the knowledge-domain it is meant to manage. Given these premises, many terms have been left in their original form to preserve their meaning integrity since the translation into Italian sometimes proved to not be necessary because of their English usage in Italian contexts, see for example *phishing*, or *honeypot*. The research has been conducted starting from a information retrieval process as the first phase, that has been meant to detect and collect the information about the domain of study following specific criteria. These parameters have been based on:

a. the range of time: the interval that has been considered is 2000–2017 for both sub-corpora, i.e., legislative and scientific;
b. the authoritativeness of the publishing sources: the trustworthy sources within the field of knowledge of Cybersercurity have been taken into account both from the legislative area regarding the regulations, norms, decrees in force about the domain of study, and from the technical and scientific sphere, ranging from the officially registered sector-oriented magazines in specific catalogues lists to the reports, guidelines and frameworks on the domain's subject issues;
c. the geographical extent: the Italian framework system on the Cybersecurity phenomena has been firstly analyzed, the English equivalent documentation has been set out to retrieve the correspondences in terminological units meant to support the process of creating multilingual semantic tools;
d. the format of the textual files: digital format have been preferred over the paper-based ones, which, where needed, have been scanned through a OCR recognition process for the scientific-popular sub-corpus. The corpus population procedure has represented the qualitative step in the coverage measurement of the knowledge-domain information: the authoritative heterogeneous updated and specifically domain-oriented complexion marking up the bilingual Cybersecurity corpora can be evaluated as a discriminating qualitative methodology in verifying the quality of the source texts. To provide a quantitative evaluation on the assembled starting corpus, two approaches have been described:

a. ratio measurement among tokens, types and hapaxes
This measurement is based on the TTR formula which divides the number of

tokens – all the words present in the source corpus – with the number of types – the unique words (1-gram; 2-gram) –, in this case is equal to 4,8%.

$$\frac{total\ number\ of\ types}{total\ number\ of\ tokens} \times 100 \qquad (7.1)$$

Despite the quite low TTR score percentage value, that stands for the terms variety in the text, the proportion of hapaxes – words occurring just once in the source documents – is an encouraging score of almost 70%. This means that the selected documents are indeed providing a specific form of unstructured information to be processed, and indicates that the corpus is qualitative ensuring a high range of terms specificity.

b. gold standards mapping system

The mapping system executed on the terms extracted with respect of the list of terms contained in the gold standards provided valuable results in the analysis of the representativeness of the corpus. The corpus candidate terms list equals **17083** units given by the terminological extraction executed with the semi-automatic software engines, and successively the thesaurus flat list of terms containing **245** terms approved by the experts of the domain, have been mapped with five selected reference frameworks (White Book [ITA], NIST [EN-translated], ISO [EN-translated], Glossary of Intelligence [ITA] and Clusit [ITA]). The analytical comparison scores computed the presence of many matches between the term extraction outputs with the list contained in the Clusit reports, and in the White Book text, namely 1701/5407 for Clusit and 2292/14635 for the White Book. This indicates a proximal level of similar semantic specificity with the selected candidate terms extraction list to become part of the thesaurus. Indeed, both Clusit report lists and White Book offer a broadly extensive description on the main episodes occurring in the field of Cybersecurity in Italian language mixing up a legislative granular information and a more technical exploration of the key concepts, as the texts in the compiled source corpus has aimed to display. When the mapping has been executed over the flat terms list of the thesaurus, which has already undergone a validation phase with the experts of the domain, with the aforementioned gold standards to demonstrate the level of semantic coverage achieved, the scores underlined that over the 245 selected terms included in the thesaurus, 80 have matched with the White Book, and this confirmed the proximity sharing the information retrieved for the Italian Cybersecurity corpus and the White Book published for the Italian Cybersecurity context, as well as for Clusit where the score indicated the match of 52 terms. Over the results obtained by intersecting the shared terminology between the source extraction list of candidate terms, as well as the subsequent thesaurus list configuration, it is worth observing the low scores with the terms contained in the Italian Glossary of Intelligence. In fact, unlike the other two reference standards in the Italian language with which the Italian Cybersecurity terminology has found a significant level of similarity, the Glossary of Intelligence has not performed the same – in the extracted candidate terms list the match is 58/266, while in the thesaurus list only 10/266 – and this is probably due to the fact that this specific glossary has been published by political organisms and it is marked by a precisely governmental-oriented language, which could be found in the more comprehensive list given by the term extractor software over the source corpus made of heterogeneous documents. A higher quantitative result has been obtained for what concerns the mapping process executed over the term lists included in NIST 7298 and ISO 27000:2016 (manually translated in Italian with the aide of the IATE platform): it can be observed that the overall term extraction series matched with NIST gave 28/446 retrieved common entries and 39/88, while in the thesaurus list only 12 for NIST mapping and 11 for ISO. The discrepancy found between the software output candidate list and the terms contained in the thesaurus is attributable both to the specification process over the terms contained in the standards that decreases the terms to be considered and proposed to the experts, for instance *file* has been not taken into account from ISO matches towards the intervention of the group of experts who validated the selected candidates. The project objective has regarded the quantitative and qualitative coverage measurement of the domain provided by two semantic sources developed for the Italian Cybersecurity framework, a thesaurus re-engineered into an ontology. The mapping of the source corpus candidate term lists resulted from the source documentation NLP processing with the software-assisted engines and the thesaurus flat term lists with the gold standards represented the first step in evaluating the semantic coverage level with respect to terminology formally shared in the state-of-the-art gold standards, the evaluation lists, of Nist, ISO, Glossary of Intelligence and Clusit. The terminological process started from the linguistic analysis of the source corpus with the help of semantic software specifically trained to extract representative semantic units. In particular, in this book the processing of the source corpus documentation carried out with the aide of T2K Italian NLP linguistic extractor tool, TermSuite terminology extraction toolbox for which Italian morpho-syntactic rules set in Java have been implemented to be run over the selected source corpus for the Italian Cybersecurity framework, Pke document oriented-library for the clustering of documents and BERT-based approach has been described. From the results obtained by Recall and Precision statistical measurement, the corpus information extraction provided by Pke has provided a higher percentage in the proportion of the correct terms retrieved by the system and all the correct terms in gold standards, i.e., semantic coverage, while BERT performed better with respect to the detection of the proportion of the all candidates detected by the system with the complex of all the correct terms in the gold standards. With regard to the term extractor tools, i.e., T2K and TermSuite, T2K provided higher percentage scores in the Recall when comparing the extractions with the NIST glossary and ISO taxonomy, namely **21.2** and **47.7**, and also for what concerns the Glossary of Intelligence and Clusit the Recall measure resulted higher than the one given by TermSuite term extraction, e.g., **23.7** for T2K extraction – Clusit and **7.92** for TermSuite extraction – Clusit. The low total number in the precision measurements is attributable to the reduced dimension (around 200) of the evaluation lists whereas the systems output are run over thousands of terms. To sum up, referring to the F1 score, BERT achieved the best results in all the cases, and T2K proved to perform better with respect to TermSuite in computing the semantic coverage to the evaluation lists, the term lists of the gold standards. Pke, which is a document-oriented library that detects with the implementation of its internal models, e.g., *TopicRank, MultipartiteRank*, the best representative keyphrases to be considered as domain-specific topics, also gave a positive output both in terms of Recall and Precision. Nonetheless, this exploratory research has highlighted the functionality of PKE mostly referring to the detection of the main macro-categories for the thesaurus' structure configuration. Thus, providing adequate percentages has represented a confirmation of how efficiently it works in recognizing relevant terms with respect to the knowledge included in the official standards. These final outputs are to be considered as statistical calculations that point out how the term extractors are functional for the the detection of the knowledge-domain representativeness level, but, on the other hand, a manual semantic evaluation demonstrated that TermSuite outputs resulted more accurate in terms of quality in the terms list provided. Therefore, if on the one hand T2K can be considered, according to the statistical computations, an efficient semantic tool with which to extract a reliable quantitative set of representative terms – considering representative also the ones retrieved as correct candidates in the officially accepted lists in the gold standards –, on the other hand, TermSuite with its internal higher specific user-adjustable pre-defined semantic rules application over languages, and its function of detecting variants in specialized source corpora, makes the toolkit software more suitable for the purposes of the semantic precision outputs achievement. To define the level of domain terms accuracy representation, firstly the principle of authoritativeness of the sources, from which the documentation is included in the corpus, has been taken into account as the first determining criterion when it comes to considering the reliability of the information provided. The publishing sources have had a relevant impact in selecting the texts to then be processed with NLP tasks. From a statistical point of view, the measures that have been considered have applied the TF*IDF formula *per* each term and the computation of the specificity level ranking.

The terms frequency has represented a preliminary discriminatory method to sort out the most representative candidate terms to undergo a validation phase by the experts of the domain. The frequency scores aimed at selecting the most indicative candidate terms towards their interrelation configuration in accordance with the standards marked up by semantic connection rules and labels (BT-NT for the hierarchies, USE-UF for the synonyms, RT for the related terms). The specificity level has been measured within the list of terms extracted by the different semi-automatic semantic tools and it has been determined in contrast with a general Italian language external corpus. This comparison process ranked terms while considering them included either in the specialized Italian Cybersecurity source corpus or in the general Italian one. This study has discussed the compliance of specific terms properly belonging to the field of Cybersecurity, as for example *malware*, *phishing*, as well as the presence of some terms occurring in the cyber contexts, such as *virus*, *security*, *control*, that showed high scores in the contrast corpus. This means that there are certain terms that are considerably frequent in the general corpus and that can indicate quite a weak threshold of specificity when it comes to start considering to be included or not as representative terms reflecting the specialized Cybersecurity knowledge-domain. even though some of them could be considered as specific of the area of study such as the aforementioned ones. In this research the enhancement of the semantic structure of the thesaurus – and consequently of the ontology –, of its relationships framework as well as their automatized detection and improvement has been investigated. In particular, the relationships that the research activities have thoroughly explored in order to guarantee an automatic enhanced systematization refer to the ones included in the standard ISO 25964-1:2011 and 25964-2:2013, namely the hierarchical structure, the synonym configuration of the preferred terms, and the associative relation occurring among several terms of the domain that are not linked together by general-specific boundaries. In this study the search for suitable techniques to address in order to improve the accuracy of the structured semantic relationships network management within the thesaurus meant to be provided in the Italian Cybersecurity area of study extent has been outlined. The approaches described have covered the (i) the head-terms detection through the aide of T2K included output configuration, according to which the most frequent terms have been treated as head-based ones followed by a list of the most specific terms with which they have occurred in the source documents, this supported the first phase of the decision-making process of the hierarchical relations; (ii) variation detection with TermSuite semi-automatic term extractor tool which has helped throughout the recognition of the hierarchical connections and synonyms in the way the morpho-syntactic rules pointed to variations in the candidate terminological units; (iii) the pattern-based configuration batteries that have represented an efficient way to recursively retrieve the indicators of the broader-specific connections among terms through the hyperonyms and meronyms rules applied detection procedure. Moreover, the patterns have been employed to discover the synonymy and the associative relations: for the first type of connection, a set of semantic structured chains helped in isolating the respective textual portions implying the equivalence relation, and for the second one a list of causative verbs patterns have supported the establishment process of the different kinds of associative relationships to be clarified. Indeed, the associative relationship given in the thesaurus is the one that causes more ambiguities in specifying the explicit connections occurring among the terms that represent the domain concepts. The causative verbal construction used to identify the type of links the terms shared has represented a key technique both for what concerns the enhanced structuring of the associative network in the thesaurus, which received before this execution the validation of the Italian Cybersecurity group of experts, and also a guideline to improve the properties to be imported in the conversion made from the thesaurus into an ontology and (iv) the word embedding models, particularly Word2Vec and FastText, proved to perform with encouraging results for the identification of new associative relations, as well as hierarchical and equivalence. Starting from the pre-established approved by experts relationships within the thesaurus for the Cybersecurity domain in the Italian language the word embedding models have supported the detection of the most proximal terms and multi-words units to associate to existing ones included in the thesaurus. The experiments underlined the effectiveness these models have provided in retrieving the list of the similar terms *per* each relation, improving, in this way, the accuracy of the semantic network of the source thesaurus. Among the confirmed relationships between the first thesaurus approved by experts configuration the following are examples:

- **verbs**
  cyber attacks [perpetrate by] hacker;
  hackers [exploit] vulnerabilities/ zero-day;

- **hyperonym**
  cyber attacks → ransomware
  ransomware → malware
  malware → virus, phishing
  cyber attacks → spoofing
  cyber attacks → DDoS
  malware → trojan, virus, worm

For what concerns the patterns procedure exploited to detect new types of term connections, the following are some of the new domain-oriented specific information that enriched the semantic tool's outline:

1. **meronymy:**
   virus → Lovebug, Code Red, Nimda, Sircam, VBS
   trojan horse → WinCrash,Sinowal (MebRoot, Torpig), Trojan downloader, Trojan dropper, Smokedown
   anonymized surfing → Anonimizer, Multiproxy
   malware → Warezov (Opnis, Stratiov), MAHADI malware, Wiper, Man in the Mobile (MitMO). Ram scaper
   attacks → mass-nuking, eavesdropping
   worm →Stuxnet
   honeynet → [type of] honeypot clockjacking → [type of] cyber attack
   Dnsspoof → [type of] cyber attack

2. **verbs**

threats [damage] privacy, integrity

blockchain [enables] cryptoccurency

Fragroute [used to bypass] recognition systems based on digital signatures

cyber insurance [covers damages] cyber terrorism

backdoor [bypasses] authentication systems

malware [installs] backdoor

worm [installs] backdoor

cyber criminals [install] malware

botnet [infected by] malicious software bot

botnet [controlled by] botmaster

antivirus [fights against] cyber threats

tunneling [bypasses] firewall

VBS [share many features with] Melissa

software ripping (SmartRipper) [bypass] protection BD technologies

worm [bypasses] antivirus

spoofing [bypasses] firewall cracker [tampers] cybersecurity mechanisms

cyber hactivism [tampers] cybersecurity

hackers [tamper] cybersecurity

ransmoware [bypasses] bitcoin

cyber attacks [encroach] cyber terrorism backdoor [used by] crackers sniffer [captures] password keylogger [informs] cracker

Guaranteeing a system of interoperability between the different informative tools that exchange information about the object of the domain has represented an important step of this present research project. The goal has been to set a broadly-expanded form of usability and reusability of the semantic tools for large and complex domain usages. The semantic information included in the thesaurus has been made exchangeably readable by different informative systems working in cyber frameworks with the implementation of two approaches:

- digitization and standardization: constructing the thesaurus in a digital form and adjusting the standardization of candidate terms in compliance with the ISO 25964-1 and 25964-2 standards;

-conversion into an ontology: to make the thesaurus' knowledge organization interoperable with other informative systems, a method that has been implemented related to its structure conversion into a conceptual ontological taxonomy by using RDF(S) and OWL languages. Indeed, the thesaurus' configuration for the Italian Cybersecurity framework is possible to export in XML and in SKOS:label syntax, and this facilitates the transposition of the terminological network in a ontology composition outline. The aim in migrating the term structure into an ontology is thus referred to the higher possibility to support the way the systems can reuse it within the standardized community of knowledge-domain expert thanks to the OWL and RDF(S) more flexible languages; and also to better represent, on a semantic level, the hierarchy and the associative connections occurring among the thesaurus' terms. In detail, RDF(S) offers the advantage with the application of its rules to define a common vocabulary towards the ontology development where the predicate entities represent a superordinate class and another entity a property of the class with which it shares a range of connections. While thanks to OWL it is possible to add extra semantic properties, e.g., the disjunction between two sets, according to which a collection of concepts can be considered separately from other ones, the functional property association to a class, i.e., the definition of the unique identifiers, the transitive property between classes, that is useful to connect some of them sharing given properties and to create a highly descriptive semantic network. The main characteristic of OWL language resides in its formalism and in the possibility to include restrictions, to provide a more formal conceptual descriptive frameworks for knowledge-domains and to apply automatic reasoners engines able to execute inferences on the field of study conceptual model. In particular this work presented several cases where the ontology, which has been developed under the basis of the Italian Cybersecurity thesaurus' structure and the enhancements applied after executing the patterns-based and word embedding models configuration over the source corpus and the pre-defined thesaurus' relations, proved to be more detailed in the way it explicitly described the interrelations existing among domain-specific concepts. This is due to the fact that in the thesaurus, even if terms are connected by applying the automatic relationships detection techniques outputs, the configured system this tool presents has to be compliant with the semantic connection structure pre-arranged by the aforementioned terminology ISO standards. Therefore, the rigidity characterizing the limited flexibility of a thesaurus to explicitly present the type of associative relation has been considered as a constraint to achieve a full-descriptive knowledge-domain representation. This is why the ontology systematization seemed to be more adequate to reach this purpose since it offers the possibility to connect the concepts by using properties that can show the exact nexus in using the verbs retrieved in the semantic relations detection methodology, pattern-based technique, making the Cybesecurity framework terminology connections in Italian language more explicit in the related associations among the representative terms.

Terminology monitoring

The research work investigated the ways by which the thesaurus and the ontology could become constant, trustworthy, monitoring resources for the domain of Cybersecurity, with the goal of expanding the scope to other fields of study whose specific terminology is meant to be explored. As a result of the methodologies applied with reference to the enhancement terms connections network, the multiple new information coming from the implementation of algorithms of word embedding as well as pattern-based queries execution have guaranteed a monitoring process in the terminological variation within the source documents. This is encouraging for new possible implementations over different specialized fields of knowledge to be managed and represented with the aide of semantic structures such as thesauri and ontologies. The perspectives within this study investigation have concerned the creation of a continuous self-integrated system detection of new cyber information to be imported in both thesaurus and ontology. It has been based on two techniques:

-SPARQL queries: by connecting to external drivers of authoritative sources documentation uploads, and starting mapping queries, it can be possible to auto-populate the ontology classes with the ones included in the reference table elements, this could allow to be in line with the updates coming from selected sources that are likely to be the same ones chosen to compile the authoritative source corpus, and, thus, keep up with the latest news about the domain under study and adjusting the semantic frame within the resources step by step;

-Twitter crawling: almost following the same principle of SPARQL queries system, the activity that can be followed up to monitor the semantic change of the domain-specific information refers to tweets analytic detection. In particular, by classifying a priori a set of trustworthy profiles publishing reliable tweets on the domain of Cybsercurity, a crawling of their posts can be executed to perform a detection process for new information identification and keep up with the modifications that can be added on the terms already included in the thesaurus and ontology. This can lead either to new terms integrations, and, thus, to more extensive constant semantic coverage of the domain, and to modifications on the current structure that can imply a re-thinking process of the existing semantic relationship configurations to best cover the terminology of the specialized area of study.

References

[1] EAGLES (1996e). *Preliminary recommendations on corpus typology.* Consiglio Nazionale delle Ricerche. Istituto di Linguistica Computazionale, Pisa.

[2] ISO/TC 46/SC 9 2013. *ISO 25964-2:2013 Information and documentation — Thesauri and interoperability with other vocabularies — Part 2: Interoperability with other vocabularies.* 2013.

[3] Adrien A., Boudin F., and Daille B. Topicrank Graph-based topic ranking for keyphrase extraction. In *Proceedings of the Sixth International Joint Conference on Natural Language Processing,* pages 543–551, Nagoya, Japan, oct 2013. Asian Federation of Natural Language Processing.

[4] Andrade D., Tsuchida M., Onishi T., and Ishikawa K. Synonym acquisition using bilingual comparable corpora. In *Proceedings of the Sixth International Joint Conference on Natural Language Processing,* pages 1077–1081, oct 2013.

[5] Auger A. and Barrière C. Pattern-based approaches to semantic relation extraction: A state-of-the-art. *Terminology,* 14:1–19, 06 2008.

[6] Aviad A., Węcel K., and Abramowicz W. The semantic approach to cyber security. towards ontology based body of knowledge. In *European Conference on Information Warfare and Security, ECCWS,* volume 2015, pages 328–336, 01 2015.

[7] Ballarin, Matteo SKOS: un sistema per l'organizzazione della conoscenza., 2006 Degree thesis thesis, University "Ca' Foscari" of Venice (Italy). [Thesis]

[8] Blumauer A. and Pellegrini T. *Semantic Web und semantische Technologien: Zentrale Begriffe und Unterscheidungen*, pages 9–25. Springer Berlin Heidelberg, Berlin, Heidelberg, 2006.

[9] Caruso A. and Folino A. *Corpus-based knowledge representation in specialized domains.*, volume 210, pages 11–36. Peter Lang, Berna, 2016.

[10] Caruso A., Folino A., Parisi F., and Trunfio R. A statistical method for minimum corpus size determination. In *12es Journées internationales d'Analyse statistique des Données Textuelles (JADT2014)*, Paris, France, 2014.

[11] Condamines A. L'interprétation en sémantique de corpus: le cas de la construction de terminologies. *Revue française de linguistique appliquée*, Vol. XII(2007/1):39–52, 2007.

[12] Condamines A. Taking genre into account when analyzing conceptual relation patterns. *Corpora*, 8:115–140, 2008.

[13] Condamines A. Terminological knowledge bases from texts to terms, from terms to texts. In *The Routledge Handbook of Lexicography*. Routledge, 2018.

[14] Erkan G., and Radev D.R. (2004). Lexrank: Graph-based lexical centrality as salience in text summarization. Journal of artificial intelligence research, 22, 457–479.

[15] Farzindar A. and Diana I. *Natural Language Processing for Social Media*. Morgan & Claypool Publishers, 2015.

[16] Ferreira A., Maculan B., and Naves M. Ranganathan and the faceted classification theory. *In Transinformação*, 29:279–295, 12 2017.

[17] Hazem A. and Daille B. Word Embedding Approach for Synonym Extraction of Multi-Word Terms. In *Proceedings of the Eleventh International Conference on Language Resources and Evaluation (LREC 2018)*, Miyazaki, Japan, May 7-12, 2018 2018. European Language Resources Association (ELRA).

[18] Miles A. and Bechhofer S. SKOS Simple Knowledge Organization System Reference. W3C Recommendation. *World Wide Web Consortium*, United States, 8 2009.

[19] Nazarenko A., Zargayouna H., Hamon O., and Van Puymbrouck J. Evaluation des outils terminologiques: enjeux, difficultés et propositions. *Traitement Automatique des Langues*, 50(1 varia):257–281, 2009.

[20] Oltramari A., Cranor L.F., Walls R.J., and McDaniel P. Building an ontology of cyber security. CEUR Workshop Proceedings, 1304:54–61, 01 2014.

[21] Ramesh A., Srinivasa K. G., and Pramod N. Sentencerank — a graph based approach to summarize text. In *The Fifth International Conference on the Applications of Digital Information and Web Technologies (ICADIWT 2014)*, pages 177–182, 2014.

[22] Rigouts Terryn A., Hoste V., and Lefever E. A Gold Standard for Multilingual Automatic Term Extraction from Comparable Corpora: Term Structure and Translation Equivalents. In *Proceedings of the Eleventh International Conference on Language Resources and Evaluation (LREC 2018)*, Miyazaki, Japan, May 7-12, 2018 2018. European Language Resources Association (ELRA).

[23] Singhal A. and Wijesekera D. Ontologies for modeling enterprise level security metrics. *In ACM International Conference Proceeding Series*, 2010.

[24] Sophie Aubin and Thierry Hamon. Improving Term Extraction with Terminological Resources. LNAI 4139, page 380. Springer, 2006.

[25] Chang A.X. and Manning C.D. TokensRegex: Defining cascaded regular expressions over tokens. Technical Report CSTR 2014-02, Department of Computer Science, Stanford University, 2014.

[26] Athiwaratkun B., Wilson A.G., and Anandkumar A. Probabilistic fasttext for multi-sense word embeddings. *In CoRR*, abs/1806.02901, abs/1806.02901.

[27] Barnett B. and Crapo A. A semantic model for cyber security. In *Implementing Interoperability, Advancing Smart Grid Standards, Architecture and Community*, Phoenix, AZ, USA, 2011.

[28] Daille B. Conceptual structuring through term variations. In F. Bond, A. Korhonen, D. MacCarthy, and A. Villacicencio, editors, *Proceedings ACL 2003 Workshop on Multiword Expressions: Analysis, Acquisition and Treatment*, pages 9–16. ACL, 2003.

[29] Daille B. Variations and application-oriented terminology engineering. *In Terminology*, 11:181–197, 2005.

[30] Daille B. *Term Variation in Specialised Corpora: Characterisation, automatic discovery and applications*. John Benjamins, 2017.

[31] Daille B. *Term Variation in Specialised Corpora: Characterisation, automatic discovery and applications*, volume 19 of *Terminology and Lexicography Research and Practice*. John Benjamins, 2017.

[32] Daille B. and Hazem A. Semi-compositional method for synonym extraction of multi-word terms. In *Proceedings of the Ninth International Conference on Language Resources and Evaluation (LREC'14)*, pages 1202–1207, Reykjavik, Iceland, May 2014. European Language Resources Association (ELRA).

[33] Fortuna B. and Grobelnik M. Semi-automatic data-driven ontology construction system. *In Information Systems - IS*, 2006.

[34] Gray B. Exploring methods for evaluating corpus representativeness. 2017.

[35] Gupta B., Aditya R., Akshay J., Arpit A., and Naresh D. Analysis of various decision tree algorithms for classification in data mining. *In International Journal of Computer Applications*, 163:15–19, 04 2017.

[36] Hjørland B. What is knowledge organization (KO)? *In Knowledge Organization. International Journal devoted to Concept Theory, Classification, Indexing and Knowledge Representation*, 2008.

[37] Richards B. Type/token ratios: what do they really tell us? *in Journal of Child Language, Cambridge University Press*, 14(2):201–209, 1987.

[38] Saini B., Singh V., and Kumar S. Information retrieval models and searching methodologies: Survey. *In International Journal of Advance Foundation and Research in Science & Engineering (IJAFRSE)*, 1, 07 2014.

[39] Wielinga B. J., Schreiber A. Th., Wielemaker J., and Sandberg J. A. C. From thesaurus to ontology. In *Proceedings of the 1st International Conference on Knowledge Capture*, K-CAP '01, page 194–201, New York, NY, USA, 2001. Association for Computing Machinery.

[40] Florian B. pke: an open source python-based keyphrase extraction toolkit. In *Proceedings of COLING 2016, the 26th International Conference on Computational Linguistics: System Demonstrations*, pages 69–73, Osaka, Japan, December 2016. The COLING 2016 Organizing Committee.

[41] Barrière C. Hierarchical refinement and representation of the causal relation. *Terminology. International Journal of Theoretical and Applied Issues in Specialized Communication*, 8(1):91–111, 2002.

[42] Barrière C. Semi-automatic corpus construction from informative texts. In Lynne Bowkes, editor, *Text-Based Studies in honour of Ingrid Meyer*, Lexicography, Terminology and Translation, chapter 5. University of Ottawa Press, January 2006.

[43] Mader C., Haslhofer B., and Isaac A. Finding quality issues in skos vocabularies. In Panayiotis Zaphiris, George Buchanan, Edie Rasmussen, and Fernando Loizides, editors, *in Theory and Practice of Digital Libraries*, pages 222–233, Berlin, Heidelberg, 2012. Springer Berlin Heidelberg.

[44] Biber D. Representativeness in Corpus Design. *In Literary and Linguistic Computing*, 8(4):243–257, 01 1993.

[45] Bourigault D. and Aussenac-Gilles N. Construction d'ontologies á partir de textes. pages 11–14, 01 2003.

[46] Cram D. and Daille B. Terminology extraction with term variant detection. In *Proceedings of ACL-2016 System Demonstrations*, pages 13–18, Berlin, Germany, August 2016. Association for Computational Linguistics.

[47] Gablasova D., Brezina V., and T. McEnery. *The Trinity Lancaster Corpus: Applications in language teaching and materials development*, pages 7–28. 2019.

[48] Ganguly D., Roy D., Mitra M., and Jones G.J.F. Word embedding based generalized language model for information retrieval. In *Association for Computing Machinery*, volume SIGIR '15, page 795–798, 2015.

[49] Kless D., Jansen L., Lindenthal J., and Wiebensohn J. A method for re-engineering a thesaurus into an ontology. In *Proceedings of International Conference on Formal Ontology in Information Systems (FOIS 2012)*, pages 133–146, 2012.

[50] Yang D. and Powers D.M. Automatic thesaurus construction. In *Proceedings of the Thirty-first Australasian Conference on Computer Science – Volume 74*, ACSC '08, pages 147–156, Darlinghurst, Australia, Australia, 2008. Australian Computer Society, Inc.

[51] S. Dang and Ahmad P. A review of text mining techniques associated with various application areas. *In International Journal of Science and Research (IJSR)*, 4 (2):2461–2466, 2015.

[52] J. Delisle, H. Lee-Jahnke, M. Ulrych, M.C. Cormier, C. Falbo, and M.T. Musacchio. *Terminologia della Traduzione*. Hoepli, 2005.

[53] Embley D.W., Liddle S.W., Lonsdale D.W., and Tijerino Y.A. Multilingual ontologies for cross-language information extraction and semantic search. In *ER*, 2011.

[54] Cardillo E., Folino A., Trunfio R., and Guarasci R. Towards the reuse of standardized thesauri into ontologies. In *Proceedings of the 5th International Conference on Ontology and Semantic Web Patterns – Volume 1302*, WOP'14, page 26–37, Aachen, DEU, 2014. CEUR-WS.org.

[55] Doynikova E., edorchenko A., and Kotenko I. Ontology of metrics for cyber security assessment. In *Proceedings of the 14th International Conference on Availability, Reliability and Security*, ARES '19, New York, NY, USA, 2019. Association for Computing Machinery.

[56] Loginova Clouet E., Gojun A., Blancafort H., Guegan M., Gornostay T., and Heid U. Reference Lists for the Evaluation of Term Extraction Tools. In *Terminology and Knowledge Engineering Conference (TKE)*, Madrid, Spain, June 2012.

[57] Morin E. and Jacquemin C. Projecting corpus-based semantic links on a thesaurus. In *Proceedings of the 37th Annual Meeting of the Association for Computational Linguistics*, page 389–396. Association for Computational Linguistics, 1999.

[58] Liddy E.D. Natural language processing. *In 2nd edn. Encyclopedia of Library and Information Science, Marcel Decker.*

[59] Baader F., Calvanese D., McGuinness D.L., Nardi D., and Patel-Schneider P.F. *The Description Logic Handbook: Theory, Implementation, and Applications.* Cambridge University Press, USA, 2003.

[60] Boudin F., Mougard H., and Cram D. How document pre-processing affects keyphrase extraction performance. *In CoRR volume abs/1610.07809*, 2016.

[61] Dell'Orletta F., Venturi G., Cimino A., and Montemagni S. T2K: a system for automatically extracting and organizing knowledge from texts. In *Proceedings of the Ninth International Conference on Language Resources and Evaluation (LREC'14)*, Reykjavik, Iceland, 2014. European Language Resources Association (ELRA).

[62] Lancaster F.W., Unesco. General Information Programme, and UNISIST (Program). *Thesaurus Construction and Use: A Condensed Course.* General Information Programme and Unisist, Unesco, 1985.

[63] Bernier-Colborne G. Defining a gold standard for the evaluation of term extractors. In Proceedings of the Eight International Conference on Language Resources and Evaluation (LREC'12), pages 15–18, 2012.

[64] Corpas Pastor G. and Seghiri M. Size matters: A quantitative approach to corpus representativeness. *in Language, translation, reception. To honor Julio César Santoyo, Universidad de León*, pages 1–35, 2010.

[65] Grefenstette G. *Explorations in Automatic Thesaurus Construction*. 1994.

[66] Hodge G. *Systems of Knowledge Organization for Digital Libraries: Beyond Traditional Authority Files*. 2000.

[67] Leech G. The state of the art in corpus linguistics. *in* Aijmer K. and Altenberg B. (eds.) *English Corpus Linguistics: Studies in Honour of JaJan Svartvik*, London: Longman, pages 8–29., 1991.

[68] Sidorov G., Miranda-Jiménez S., Viveros-Jiménez F., Gelbukh A., Castro-Sánchez N., Velásquez F., Díaz-Rangel I., Suárez-Guerra S., and Gordon A., Treviño andJ. Empirical study of machine learning based approach for opinion mining in tweets. In Ildar B. and Miguel G. M., (eds.). *Advances in Artificial Intelligence*, pages 1–14, Berlin, Heidelberg, 2013. Springer Berlin Heidelberg.

[69] Thurmair G. Making term extraction tools usable. In *EAMT Workshop: Improving MT through other language technology tools: resources and tools for building MT*, Budapest, Hungary, April 13 2003. European Association for Machine Translation.

[70] Virginia G. and Nguyen H.S. Investigating the effectiveness of thesaurus generated using tolerance rough set model. In Marzena K., Henryk R., Andrzej S., and Zbigniew W. R., (eds.), *Foundations of Intelligent Systems*, pages 705–714, Berlin, Heidelberg, 2011. Springer Berlin Heidelberg.

[71] Zagrebelsky G. *Il sistema costituzionale delle fonti del diritto*. UTET, Turin, 1984.

[72] Cea G. A. D. and Montiel-Ponsoda E. Term variants in ontologies. 2012.

[73] Venkat N. Gudivada, Dhana L. Rao, and Amogh R. Gudivada. Chapter 11 - information retrieval: Concepts, models, and systems. In Venkat N. G. and C.R. Rao, (eds), *Computational Analysis and Understanding of Natural Languages: Principles, Methods and Applications*, volume 38 of *Handbook of Statistics*, pages 331–401. Elsevier, 2018.

[74] Jelodar H., Wang Y., Yuan C., Feng X., Jiang X., Li Y., and Zhao L. Latent dirichlet allocation (lda) and topic modeling: models, applications, a survey, 2018.

[75] Lesca H. and Rouibah K. Des outils au service de la veille stratégique. *In French Journal of Management Information Systems*, 2:101–131, 1997.

[76] Nakagawa H. Automatic term recognition based on statistics of compound nouns. *Terminology. International Journal of Theoretical and Applied Issues in Specialized Communication*, 6(2):195–210, 2000.

[77] Wu H. and Zhou M. Optimizing synonym extraction using monolingual and bilingual resources. In *In Proceedings of the second international workshop on Paraphrasing*, page 72, 2003.

[78] Toru Hisamitsu, Yoshiki Niwa, and Jun'ichi Tsujii. A method of measuring term representativeness - baseline method using co-occurrence distribution. In *COLING*, 2000.

[79] Meyer I. Extracting knowledge-rich contexts for terminography: A conceptual and methodological framework. In *Recent Advances in Computational Terminology*, pages 279–302. John Benjamins, 2001.

[80] Roesiger I., Bettinger J., Schäfer J., Dorna M., and Heid U. Acquisition of semantic relations between terms: how far can we get with standard NLP tools? In *Proceedings of the 5th International Workshop on Computational Terminology (Computerm 2016)*, pages 41–51, Osaka, Japan, December 2016. The COLING 2016 Organizing Committee.

[81] Nagy I.K. English for special purposes: Specialized languages and problems of terminology.*In Acta Universitatis Sapientiae, Philologica*, 6(2):261–273, 2015.

[82] ISO/IEC 27000. *Information technology — Security techniques — Information security management systems — Overview and vocabulary*. February 2016.

[83] ISO/TC 46/SC 9. *ISO 25964-1:2011 Information and documentation — Thesauri and interoperability with other vocabularies — Part 1: Thesauri for information retrieval*. International Standard, August 2011.

[84] Devlin J., Chang M., Lee K., and Toutanova K. Bert: Pre-training of deep bidirectional transformers for language understanding. In *NAACL-HLT*, 2019.

[85] Francom J., LaCross A., and Ussishkin A. How specialized are specialized corpora? behavioral evaluation of corpus representativeness for Maltese. In *Proceedings of the Seventh International Conference on Language Resources and Evaluation (LREC'10)*, Valletta, Malta, May 2010. European Language Resources Association (ELRA).

[86] Freixa J. Causes of denominative variation in terminology: A typology proposal. *In Terminology. International Journal of Theoretical and Applied Issues in Specialized Communication*, 12(1):51–77, 2006.

[87] Pearson J. *Terms in Context*. John Benjamins, Amsterdam, 1998.

[88] Qin J. and Paling S. Converting a controlled vocabulary into an ontology: the case of gem. *Inf. Res.*, 6, 2001.

[89] Sankaranarayanan J., Samet H., Teitler B.E., Lieberman M.D., and Sperling J. Twitterstand: News in tweets. In *Proceedings of the 17th ACM SIGSPATIAL International Conference on Advances in Geographic Information Systems*, GIS '09, page 42–51, New York, NY, USA, 2009. Association for Computing Machinery.

[90] Smedt J., Isaac A., Dextre Clarke S., Lindenthal J., Zeng M., Tudhope D., Will L., and Vladimir V. *ISO 25964 Part 1: Thesauri for information retrieval: RDF/OWL vocabulary, extension of SKOS and SKOS-XL*. 2013.

[91] Vivaldi J. and Rodríguez H. Evaluation of terms and term extraction systems: A practical approach. *In Terminology. International Journal of Theoretical and Applied Issues in Specialized Communication*, 13(2):225–248, 2007.

[92] Dancette J. E. and L'Homme M.C. Building specialized dictionaries using lexical functions. *In Linguistica Antverpiensia, New Series – Themes in Translation Studies*, 3:113–131, 2004.

[93] Rowley J. E. and Hartley R. J. Organizing knowledge: an introduction to managing access to information/Jennifer Rowley and Richard Hartley. Ashgate Aldershot, England ; Burlington, VT, 4th ed. edition, 2008.

[94] Bernard J. Jansen, Mimi Zhang, Kate Sobel, and Abdur Chowdury. Twitter power: Tweets as electronic word of mouth. *J. Am. Soc. Inf. Sci. Technol.*, 60(11):2169–2188, November 2009.

[95] Sager J.C. *A Practical Course in Terminology Processing*. John Benjamins, 1990.

[96] Frantzi K., Ananiadou S., and Mima H. Automatic recognition of multiword terms: The c-value/ nc-value method. In *Int. J. on Digital Libraries*, volume 3, pages 115–130, 08 2000.

[97] Kageura K. and Umino B. Methods of automatic term recognition: a review. *Terminology*, 3(2):259–289, 1996.

[98] Kageura K., Tsuji K., and Aizawa A.N. Automatic thesaurus generation through multiple filtering. In *Proceedings of the 18th Conference on Computational Linguistics – Volume 1*, COLING '00, page 397–403, USA, 2000. Association for Computational Linguistics.

[99] Kettunen K. Can type-token ratio be used to show morphological complexity of languages? *in Journal of Quantitative Linguistics*, 21(3):223–245, 2014.

[100] Stratos K. Mutual information maximization for simple and accurate part-of-speech induction. In *Proceedings of the 2019 Conference of the North American Chapter of the Association for Computational Linguistics: Human Language Technologies, Volume 1 (Long and Short Papers)*, pages 1095–1104, Minneapolis, Minnesota, June 2019. Association for Computational Linguistics.

[101] Weller K. What do we get from twitter – and what not? a close look at twitter research in the social sciences. *In Knowledge Organization*, 41(3):1–15, 2014.

[102] R. Kisserl. *Glossary of Key Information Security Terms.* National Institute of Standards and Technology, May 2013. NISTIR 7298 Revision 2.

[103] Olivier Kraif. Qu'attendre de l'alignement de corpus multilingues ? 01 2006.

[104] Hasan K.S. and Ng V. Automatic keyphrase extraction: A survey of the state of the art. In *Proceedings of the 52nd Annual Meeting of the Association for Computational Linguistics (Volume 1: Long Papers)*, pages 1262–1273, Baltimore, Maryland, June 2014. Association for Computational Linguistics.

[105] Matt J. Kusner, Yu Sun, Nicholas I. Kolkin, and Kilian Q. Weinberger. From word embeddings to document distances. In *Proceedings of the 32nd International Conference on International Conference on Machine Learning – Volume 37*, ICML'15, page 957–966. JMLR.org, 2015.

[106] Bowker L. and Pearson J. *Working with Specialized Language: A Practical Guide to Using Corpora.* London/New York: Routledge, 2002.

[107] Dekang L. Automatic retrieval and clustering of similar words. In *Proceedings of the 36th Annual Meeting of the Association for Computational Linguistics and 17th International Conference on Computational Linguistics - Volume 2*, ACL '98, pages 768–774, Stroudsburg, PA, USA, 1998. Association for Computational Linguistics.

[108] Lefeuvre L. *Analyse des marqueurs de relations conceptuelles en corpus spécialisé: recensement, évaluation et caractérisation en fonction du domaine et du genre textuel.* PhD thesis, 2017. Thèse de doctorat dirigée par Condamines, Anne et Rebeyrolle, Josette Sciences du langage Toulouse 2 2017.

[109] Lefeuvre L. and Condamines A. Constitution d'une base bilingue de marqueurs de relations conceptuelles pour l'élaboration de ressources termino-ontologiques. In Faber P. and Poibeau T., (eds.), *Terminology and Artificial Intelligence (TIA'2015)*, pages 183–190, Granada, Spain, 2015.

[110] Obrst L., Chase P., and Markeloff R. Developing an ontology of the cyber security domain. In STIDS, 2012.

[111] Van der Plas L. and Tiedemann J. Finding synonyms using automatic word alignment and measures of distributional similarity. In *21st International Conference on Computational Linguistics and 44th Annual Meeting of the Association for Computational Linguistics ACL'06*, Sydney, Australia, 2006.

[112] Wang L. *Support Vector Machines: Theory and Applications (Studies in Fuzziness and Soft Computing)*. Springer-Verlag, Berlin, Heidelberg, 2005.

[113] Marie-Claude L'Homme. *Terminologies and taxonomies*. 2015.

[114] A. C. Liang, Lauser B., and M. Sini. From agrovoc to the agricultural ontology service/concept server – an owl model for creating ontologies in the agricultural domain. In *OWLED*, 2006.

[115] Cabré Castellví M. El principio de poliedricidad: la articulación de lo discursivo, lo cognitivo y lo lingüístico en terminología (i). *In Ibérica: Revista de la Asociación Europea de Lenguas para Fines Específicos (AELFE), ISSN 1139-7241, Nº. 16, 2008, pags. 9-36*, 16, 10 2008.

[116] Hagiwara M. A supervised learning approach to automatic synonym identification based on distributional features. In *Proceedings of the ACL-08: HLT Student Research Workshop*, pages 1–6, Columbus, Ohio, June 2008. Association for Computational Linguistics.

[117] Maheswari M. Text mining: Survey on techniques and applications. *In International Journal of Science and Research (IJSR)*, 6, 06 2017.

[118] Nowroozi M., Mirzabeigi M., and Sotudeh H. The comparison of thesaurus and ontology: Case of asis&t web-based thesaurus and designed ontology. Library Hi Tech, 36, 01 2018.

[119] Rennesson M., Georget M., Paillard C., Perrin O., Pigeotte H., and Tête C. Le thésaurus, un vocabulaire contrôlé pour parler le même langage. *Médecine Palliative*, 2020.

[120] Van Assem M., Menken R. M., Schreiber G., Wielemaker J., and Wielinga B. A method for converting thesauri to rdf/owl. In Sheila A. McIlraith, Dimitris Plexousakis, and Frank van Harmelen, editors, *in The Semantic Web – ISWC 2004*, pages 17–31, Berlin, Heidelberg, 2004. Springer Berlin Heidelberg.

[121] Van Assem M., Malais V., Miles A., and Schreiber G. A method to convert thesauri to skos. pages 95–109, 06 2006.

[122] Van Assem M., Malaisé V., Miles A., and Schreiber G. A method to convert thesauri to skos. In York Sure and John Domingue, editors, *The Semantic Web: Research and Applications*, pages 95–109, Berlin, Heidelberg, 2006. Springer Berlin Heidelberg.

[123] Van Assem M., Miles v., Malaisé A., and Schreiber G. A method to convert thesauri to skos. pages 95–109, 06 2006.

[124] Vàzquez M. and Antoni O. Improving term candidates selection using terminological tokens. *Terminology. International Journal of Theoretical and Applied Issues in Specialized Communication*, 24(1):122–147, 2018.

[125] Weller M., Gojun A., Heid U., Daille B., and Harastani R. Simple methods for dealing with term variation and term alignment. In *9th International Conference on Terminology and Artificial Intelligence (TIA 2011)*, pages 87–93, Paris, France, November 2011.

[126] Hearst M. A. Automatic acquisition of hyponyms from large text corpora. In *COLING 1992 Volume 2: The 15th International Conference on Computational Linguistics*, 1992.

[127] Zeng M. L. and Mayr P. Knowledge organization systems (KOS) in the semantic web: A multi-dimensional review. *CoRR*, abs/1801.04479, 2018.

[128] Cabré Castellví M. T., Estopà Bagot R., and Vivaldi Palatresi J. Automatic term detection: A review of current systems. 2001.

[129] Tamsin Maxwell. *PhD thesis: Term Selection in Information Retrieval*. PhD thesis, 01 2014.

[130] Parmelee M.C. Toward an ontology architecture for cyber-security standards. In *STIDS*, 2010.

[131] Jockers M.L. and Thalken R. Hapax Richness, pages 93–97. *in Text Analysis with R: For Students of Literature*, Springer International Publishing, Cham, 2020.

[132] Zeng M.L. and Mayr P. Knowledge organization systems (KOS) in the semantic web: A multi-dimensional review. *In CoRR*, abs/1801.04479, 2018.

[133] Cabré M.T. La teoría comunicativa de la terminología, una aproximación lingüística a los términos. *In Revue française de linguistique appliquée*, XIV:9–15, 2009/2.

[134] Pazienza M.T., Pennacchiotti M., and F.M. Zanzotto. Terminology extraction: An analysis of linguistic and statistical approaches. In Spiros Sirmakessis, editor, *Knowledge Mining*, pages 255–279, Berlin, Heidelberg, 2005. Springer Berlin Heidelberg.

[135] Aussenac-Gilles N., Desprès S., and Szulman S. The terminae method and platform for ontology engineering from texts. In *Ontology Learning and Population*, 2008.

[136] Bel N. Corpus representativeness for syntactic information acquisition. In *Proceedings of the ACL Interactive Poster and Demonstration Sessions*, pages 138–141, Barcelona, Spain, July 2004. Association for Computational Linguistics.

[137] Guarino N., Oberle D., and Staab S. What is an ontology? In *Handbook on Ontologies*, pages 1–17. Springer, Berlin, Heidelberg, 05 2009.

[138] Shadbolt N., O'Hara K., and L. Crow. The experimental evaluation of knowledge acquisition techniques and methods: history, problems and new directions. *In International Journal of Human-Computer Studies*, 51(4):729 – 755, 1999.

[139] B. V. L. Narayana and Sreenivasa P. Kumar. A new clustering technique on text in sentence for text mining. 2015.

[140] Noy N.F. and Mcguinness D. L. Ontology development 101: A guide to creating your first ontology. Technical report, 2001.

[141] Eric Nguyen. Text mining and network analysis of digital libraries in R. In Yanchang Z. and Yonghua C., editors, *In Data Mining Applications with R*, pages 95 – 115. Academic Press, Boston, 2014.

[142] Bojanowski P., Grave E., Joulin A., and Mikolov T. Enriching word vectors with subword information. *Transactions of the Association for Computational Linguistics*, 5:135–146, 2017.

[143] Dury P. and Drouin P. L'obsolescence des termes en langues de spécialité: une étude semi-automatique de la nécrologie en corpus informatisés, appliquée au domaine de l'écologie. pages 1–11. Aarhus University, Department of Business Communication, 2010.

[144] Shotorbani P.Y., Ameri F., Kulvatunyou B., and Ivezic N. A Hybrid Method for Manufacturing Text Mining Based on Document Clustering and Topic Modeling Techniques. In *IFIP International Conference on Advances in Production Management Systems (APMS)*, volume AICT-488 of *Advances in Production Management Systems. Initiatives for a Sustainable World*, pages 777–786, Iguassu Falls, Brazil, September 2016. Part 18: Service-Oriented Architecture for Smart Manufacturing System, an SM & CPPS SIG Workshop Session.

[145] Behrang Q. Z. and Handschuh S. The ACL RD-TEC: A dataset for benchmarking terminology extraction and classification in computational linguistics. In *Proceedings of the 4th International Workshop on Computational Terminology (Computerm)*, pages 52–63, Dublin, Ireland, August 2014. Association for Computational Linguistics and Dublin City University.

[146] Albertoni R., De Martino M., and Quarati A. Integrated quality assessment of linked thesauri for the environment. In Andrea Kő and Enrico Francesconi, editors, *Electronic Government and the Information Systems Perspective*, pages 221–235, Cham, 2016. Springer International Publishing.

[147] Baldoni R., De Nicola R., and Prinetto P. Il Futuro della Cybersecurity in Italia: Ambiti Progettuali Strategici Progetti e Azioni per difendere al meglio il Paese dagli attacchi informatici. Laboratorio Nazionale di Cybersecurity (CINI) - Consorzio Interuniversitario Nazionale per l'Informatica, 2018.

[148] Davis R., Shrobe H., and Szolovits P. What is a knowledge representation? *In AI Magazine*, 14:17, 03 2002.

[149] Girju R., Badulescu A., and Moldovan D. Automatic discovery of part-whole relations. *Computational Linguistics*, 32(1):83–135, 2006.

[150] Miyata R. and Kageura K. Building controlled bilingual terminologies for the municipal domain and evaluating them using a coverage estimation approach. *In Terminology. International Journal of Theoretical and Applied Issues in Specialized Communication*, 24(2):149–180, 2018.

[151] Rapp R. Using collections of human language intuitions to measure corpus representativeness. In *Proceedings of COLING 2014, the 25th International Conference on Computational Linguistics: Technical Papers*, pages 2117–2128, Dublin, Ireland, August 2014. Dublin City University and Association for Computational Linguistics.

[152] Rocha Souza R., Tudhope D., and Almeida M. Towards a taxonomy of kos: Dimensions for classifying knowledge organization systems. *In KNOWLEDGE ORGANIZATION*, 39:179–192, 01 2012.

[153] Talib R., M., Kashif, Ayesha S., and Fatima F. Text mining: Techniques, applications and issues. *In International Journal of Advanced Computer Science and Applications*, 7, 11 2016.

[154] Arora S., Yingyu L., and Tengyu M. A simple but tough to beat baseline for sentence embeddings. In *Proceedings of the 17th International Conference on Learning Representations (ICLR'17)*, pages 1–11, 2017.

[155] Hunston S. Corpora in applied linguistics. *Cambridge Applied Linguistics, Cambridge University Press*, 2002.

[156] Pavel S., Canada. Bureau de la traduction. Direction de la terminologie et de la normalisation, D. Nolet, and Canada. Travaux publics et services gouvernementaux Canada. Direction de la terminologie et de la normalisation. *Précis de terminologie*. Terminologie et normalisation, Bureau de la traduction, 2002.

[157] Staab S. and Studer R. *Handbook on Ontologies (International Handbooks on Information Systems)*. SpringerVerlag, 2004.

[158] Wong S. and Yao Y. An information-theoretic measure of term specificity. *In J. Am. Soc. Inf. Sci.*, 43:54–61, 1992.

[159] Gries S. Th. Dispersions and adjusted frequencies in corpora: further explorations. In *Corpus-linguistic applications*, page 197–212, 2010.

[160] Salloum S.A., Al-Emran M., Monem A., and Shaalan K. Using Text Mining Techniques for Extracting Information from Research Articles, pages 373–397. Springer International Publishing, Cham, 2018.

[161] C.K. Schultz, R.H. Orr, and P.B. Henderson. *Evaluation of Indexing by Group Consensus: Final Report*. ED 025 288. Institute for Advancement of Medical Communication, 1968.

[162] Ali Asghar Shiri and Crawford Revie. Thesauri on the web: current developments and trends. Online Information Review, 24(4):273–280, 2000.

[163] Dagobert Soergel. Multilingual thesauri and ontologies in cross-language retrieval. 1997.

[164] Dagobert Soergel. Knowledge organization systems. overview. *In Bates ELIS*, 2009.

[165] Karen Sparck Jones. *A Statistical Interpretation of Term Specificity and Its Application in Retrieval*, page 132–142. Taylor Graham Publishing, GBR, 1988.

[166] Mikolov T., Yih S.W., and Zweig G. Linguistic regularities in continuous space word representations. In *Proceedings of the 2013 Conference of the North American Chapter of the Association for Computational Linguistics: Human Language Technologies (NAACL-HLT-2013)*. Association for Computational Linguistics, May 2013.

[167] Mondary T., Després S., Nazarenko A., and Szulman S. Construction d'ontologies à partir de textes : la phase de conceptualisation. In *19èmes Journées Francophones d'Ingénierie des Connaissances (IC 2008)*, pages 87–98, Nancy, France, June 2008.

[168] Schnabel T., Labutov I., Mimno D., and Joachims T. Evaluation methods for unsupervised word embeddings. In *Proceedings of the 2015 Conference on Empirical Methods in Natural Language Processing*, pages 298–307, Lisbon, Portugal, September 2015. Association for Computational Linguistics.

[169] Takahashi T. and Kadobayashi Y. Reference ontology for cybersecurity operational information. *In The Computer Journal*, 58(10):2297–2312, 2015.

[170] Gruber T.R. A translation approach to portable ontology specifications. *In Knowledge Acquisition*, 5(2):199 – 220, 1993.

[171] Broughton V. *Costruire Thesauri: strumenti per indicizzazione e metadati semantici.* EditriceBibliografica, 2008, Milano, Italia Cliffs, NJ, 2008.

[172] Kosa V., Chaves-Fraga D., Naumenko D., Yuschenko E., Badenes-Olmedo C., Ermolayev V., and Birukou A. Cross-evaluation of automated term extraction tools by measuring terminological saturation. In *Information and Communication Technologies in Education, Research, and Industrial Applications*, pages 135–163, Cham, 2018. Springer International Publishing.

[173] Raghavan V., Bollmann P., and Jung G. S.. A critical investigation of recall and precision as measures of retrieval system performance. *ACM Trans. Inf. Syst.*, 7(3):205–229, July 1989.

[174] Blondel V.D. and Senellart P. Automatic extraction of synonyms in a dictionary. *SIAM Workshop on Text Mining*, 2002.

[175] Zhang X., Huang H., and Zhang K. Knn text categorization algorithm based on semantic centre. In *2009 International Conference on Information Technology and Computer Science*, volume 1, pages 249–252, 2009.

[176] Bengio Y., Ducharme R., Vincent P., and Janvin C. A neural probabilistic language model. *In J. Mach. Learn. Res.*, 3:1137–1155, 2003.

[177] Gallina Y., Boudin F., and Daille B. Large-scale evaluation of keyphrase extraction models. In *Proceedings of the ACM/IEEE Joint Conference on Digital Libraries in 2020*, JCDL '20, page 271–278, New York, NY, USA, 2020. Association for Computing Machinery.

[178] Goldberg Y. and Levy O. word2vec explained: deriving mikolov et al.'s negative-sampling word-embedding method. *In CoRR*, abs/1402.3722abs/1402.3722, 2014.

[179] Rinott Y. On two-stage selection procedures and related probability-inequalities. *In Communications in Statistics - Theory and Methods*, 7(8):799–811, 1978.

[180] Yaakov Y. Segmentation of expository texts by hierarchical agglomerative clustering. *In CoRR*, cmp-lg/9709015, 1997.

[181] Ballarin, Matteo SKOS: un sistema per l'organizzazione della conoscenza., 2006 Degree thesis, University "Ca' Foscari" of Venice (Italy). [Thesis].

[182] Hagiwara, Masato. "A Supervised Learning Approach to Automatic Synonym Identification Based on Distributional Features." ACL (2008).

Index

A

AGROVOC, 6, 26, 49, 112, 157
Alinea, 128
Altalex, 29, 65
AntConc, 70
ATTCK, 28
Automatic Term Extraction (ATE), v, 17, 149

B

Bidirectional Encoder Representations from Transformers (BERT), 88–90, 139, 140, 154
BNCF, ix, 63–65

C

Categorization, 1, 12, 13, 37, 50, 53, 88, 132, 162
CINI Consortium, xv, 85, 132, 160
Classification Reserarch Group (CRG), 44
CLUSIT, xi, 32, 67, 86–88, 90, 92, 132, 138–140
Clustering, xxi, 12, 17, 45, 51, 78, 81, 83, 139, 156, 159, 163
CNC-Value, 19
CoE Electronic Evidence Guide, 67
Colon Classification, 44

Committee for the Security oF the Republic (CISR), 31
Common Vulnerabilities and Exposures (CVE), 28
Comparable corpus, 32
Computational Linguistics, ILC Institution, 71, 150–152, 155, 156–162
Computer Emergency Response Team (CERT), xv, 30, 33, 65, 67
Computer Security Incident Response Team (CSIRT), 30
Continuous Bag of Words (CBOW), 99
Controlled vocabulary, i, xiii, xvi, 16, 17, 36, 40, 46, 105, 108, 114, 154
Corpus compilation, 2, 12, 61, 64, 69
Corpus design, v, xi, 2, 10, 11, 29, 151
Corpus linguistics, v, xx, 1, 3, 4, 10, 153
Corpus representativeness, v, ix, 1, 2, 3, 4, 21, 150, 152, 154, 159, 160
Corpus size, xi, 8, 68, 148
Coverage, i, xiii, xvii, xviii, 2–9, 11, 15, 19, 31–33, 39, 56, 63, 65, 81, 86, 88, 90, 106, 127, 128, 130, 132, 134–140, 160

Cyber Intelligence and Information
    Security – CIS Sapienza
    Italian Cybersecurity Report,
    67
Cybersecurity, v, vi, ix, xi, xiii, xv, xvi,
    xviii–xx, 7–10, 12, 14–18,
    20, 23–30, 31–36, 38, 40, 42,
    44, 45, 46, 48–54, 56, 57,
    61–64, 67–70, 72–76, 78–80,
    82–88, 89–95, 96–102,
    103–109, 110–116, 117–123,
    124, 125, 127–133, 134–140,
    141–144, 160, 162
Cybersecurity Observatory[CE1],
    OCS, v, ix, xviii, 23–25, 29,
    34, 36, 91, 105, 115, 135,
    136
CybersecurityLab, xix

D
Dbpedia, 6
Decision Trees, 13
DeJure, 65
Description Logic (DL), 45, 48, 152
Diritto dell'Informazione e
    dell'Informatica, 65, 67
Distributional semantics, xiii, xvii, 27,
    51
DOLCE ontology, 49
Domain corpus, 13, 19

E
EUR-lex, 29, 65
European Union Agency for
    Cybersecurity (ENISA), 30,
    31, 65, 132, 133
European Unionon of Terminology
    termbanks system (IATE),
    85, 128, 139
Eurovoc, 6
Extensible Markup Language (XML),
    46, 70, 114, 128, 143

F
F1-measure, F1 measure, F1, xi, 20,
    88, 90, 140
FastText, xi, 51, 56, 94, 99, 142, 149
Food and Agricultural Organization
    (FAO), 26

G
General Data Protection Regulation
    (GDPR), 23, 29
Glossary of Intelligence, xi, xv, 33, 34,
    63, 64, 83, 85, 88, 93, 110,
    112, 138, 139, 156
GNOSIS, Italian Intelligence
    Magazine, 65, 67, 68
Gold standard, v, vi, xi, xvi, xix, xx, 1,
    31, 34, 63, 83, 85, 86, 88,
    90–92, 106, 108, 116, 124,
    138–140, 149, 152
Granularity, 3, 16, 83

H
Hacker Journal, 65, 67, 68
Hapax, xi, 8, 69, 137, 138, 158

I
Indexer, xvii, 26, 40, 42, 48
Indexing, 17, 38, 39, 46, 48–50, 53,
    71, 150, 161
Informatica e Diritto, 65, 67
Information and Computing
    Technologies (ICT), xiii, xxi,
    67, 102, 108, 128, 135, 144,
    159
Information Retrieval (IR), xix, 1, 5, 6,
    14, 17, 29, 38, 39, 51, 62,
    124, 137, 150, 151, 153–155,
    158
Institute of Information and
    Telematics (IIT), 23, 24, 29,
    91, 108, 123, 135

Interoperability, xvi, xvii, 7, 24, 29, 38, 39, 46, 47, 49, 58, 143, 147, 149, 154
Inverse Document Frequency (IDF), 5, 19, 53, 54, 84
ISACA, 35
ISO 25964:2011, ISO 25964–1, ISO 25964–1:2011, xvii, 39, 41, 49, 72, 94, 104, 141, 143, 154, 155
ISO 25964–2:2013, ISO 25964–2, ISO 25964:2013, 16, 43, 44, 48, 49, 118, 141, 143, 153
ISO 27000:2016, xxi, 17, 24, 34, 83, 112, 139
ISO/TC 46/SC 2013, 16, 153, 154
ITASEC, 132

J
JAVA, 74, 75, 82, 94
JStor, 63

K
Karlsruhe Institute of Technologies database, 63
Keyphrases extraction, vi, 51, 52, 53, 56, 147, 150, 152, 156, 162
K-means, 12, 46
K-nearest Neighbor Classification (KNN), 13, 162
Knowledge engineering, xiii, 1, 16, 17, 27, 31, 152, 162
Knowledge Organization Systems (KOSs), vi, ix, xx, 37, 38, 39, 47, 50, 124, 135, 143, 148, 158, 160, 161

L
LancsBox, 70
Latent Dirichlet Algorithm (LDA), 50, 51, 56, 79, 153

Latent Semantic Analysis LSA, 56
Loglikelihood, 19

M
Medical Subject Heading (MeSH), ix, 6, 26, 27, 49, 112
MITRE, 28
MultipartiteRank, MPRank, 51, 52, 79, 140
Mutual information, 18, 19, 156

N
Named Entity Recognition (NER), 12
National Council of Research, CNR, 23, 24, 29, 71, 91, 108, 127, 135
National Cybercrime Centre for Critical Infrastrucutre Protec-tion (CNAIPIC), 31
National Framework for Cybersecurity and Data Protection, 33
National Initiative for Cybersecurity Careers and Studies (NICCS), 34
National Institute of Standards and Technology, NISTIR 7298 NIST, Nist 7298, xxi, 17, 24, 28, 33, 34, 83, 85, 86, 88, 90, 92, 110, 132, 133, 138, 138, 156
Natural Language Processing (NLP), xv, xvi, xviii, xix, 1, 4, 12, 17, 19, 45, 54, 56, 67, 69, 147, 148, 152, 162
N-gram, xxi, 70, 71
NISO TR-06–2017, 16
Nooj, 70

O
OCLC WorldCat database, 63
OntoGen, 46

Ontology, vii, ix-xi, xiii, xv-xxi, 7–10,
15, 17, 19, 20, 21, 23, 24, 25,
31, 35, 36, 42, 43–50, 57, 80,
81, 84, 88, 91, 94, 105, 107,
112, 115–121, 123–125, 127,
128, 130–133, 135, 136, 139,
141–145, 147, 149–154,
157–159, 161, 162
OPAC, 63
Osservatorio Attacchi applicativi in
Italia Report (OAD), 67
OWL, xvi, xvii, xxi, 24, 43, 46–50, 59,
91, 105, 107, 112, 114, 115,
116, 118, 121, 124, 128, 143,
144, 155, 157, 160

P
Parallel corpus, 31, 67
Patterns, xxi, 10, 18, 27, 44, 54, 57,
58, 59, 70, 73, 75, 77, 78,
94–96, 98, 115, 117, 120,
123, 141, 142, 148, 152
Personality, Matter, Energy, Space,
Time (PMEST), 44
Pisa, xix
PKE, 50, 52, 79, 80, 81, 90, 139, 140,
150
PosTagging (POS), Part-of-Speech
tagging, 13, 53
Precision and Recall, xi, xix, 5, 53, 88,
90, 99, 128, 139, 140, 162
Pre-processing, v, 1, 13, 16, 74, 151
Protégé, x, xi, 45, 115, 117, 119
Python, xix, 52, 79, 150

R
Reference corpus, 9, 16, 129
Registro degli operatori di
comunicazione – ROC
register (AGCOM), 63
Resource Description Framework
(RDF), xvi, xvii, xxi, 10, 21,
24, 46, 48, 49, 50, 107, 112,
114, 115, 117, 118, 124, 129,
130, 143–145

S
Security Intelligence Department
(DIS), 30
Semantic relationship, Semantic
relation, vi, ix, x, xvii, xix, 9,
16, 19, 26, 38–41, 44, 51, 53,
54, 56, 57, 61, 62, 72, 75, 79,
90, 93, 94, 99, 100, 104, 105,
107, 108, 114, 118, 119, 121,
123, 130, 131, 136, 137, 141,
144, 145, 147, 154
Semantic representativeness, 5
Semantic tool, vi, xv, xvii, xviii, xix,
2, 5–10, 17, 19, 24–26, 34,
36, 37, 44, 62, 105–107, 118,
123–125, 127, 128, 131, 132,
136, 137, 140–143
Semantic variation, xxi, 55
Simple Knowledge Organization
System (SKOS), vi, 6, 39,
49, 50, 143, 147, 148, 151,
155, 157, 158
Source corpus, xi, xv, xix, 5, 7, 10, 15,
17, 21, 27, 29–34, 43, 50, 52,
54, 61, 62, 65, 66, 69, 73–75,
79, 82, 84, 86, 89, 93–95, 98,
99, 106, 107, 117, 124
SPARQL, xix, xx, 10, 107, 115, 131,
145
Specialized corpus, 2, 5, 11, 16–18,
21, 57, 69
Specialized domain, v, xv, xvii, 3–6, 9,
16–20, 25, 26, 39, 41, 50, 53,
90, 99, 106, 112, 125, 148
Specialized languages, v, 17, 25, 26,
50, 154, 156
Specificity, 2, 3, 6, 14, 16, 20, 21, 26,
29, 31, 42, 43, 66, 69, 71, 73,
75, 84, 106, 128, 130, 138,
140, 141, 161

Standardization, 20, 114, 124, 125, 136, 143
Stanford TokensRegex, 75, 149
Stemming, 6, 74
Support Vector Machine (SVM), 13, 46, 157

T
Taxonomy, 26, 34, 39, 45, 83–85, 91, 136, 139, 143, 160
Tematres, 105
Term Frequency (TF), 19, 53, 54, 84
Term Frequency- Inverse Document Frequency (TF/IDF), Tf*IDF, TF x IDF, tf-idf, tf*idf, 19–21, 51–54, 79, 84, 140
Termhood, ix, 5, 20, 74, 75, 84, 85
Terminae, 46, 159
Terminological extraction, Term extraction, vi, xi, xix, xx, xxi, 14, 17–19, 32, 41, 46, 53, 69, 71, 72, 83, 86, 90–92, 138–140, 149, 152, 153, 155, 162
Terminological Knowledge Bases (TBKs), v, xvi, 12, 15, 16, 148
Terminological representativeness, Term representativeness, v, vii, 5, 9, 129, 154
Terminology, xix, 2, 5, 7–11, 14, 15, 17, 18, 20, 24, 26, 28, 31, 33–37, 39, 40, 42, 49, 51, 52, 54–58, 66, 67, 70–73, 75, 76, 79–81, 83–86, 88, 90, 91, 93, 99, 106–108, 112, 115, 117, 121, 123–125, 127, 128, 130, 132, 133, 135, 136, 138, 139, 144, 145, 147, 149–156, 158, 160

Termsuite, vi, xi, 71, 73–75, 78, 82–84, 86, 90, 93, 94, 139–141
Text mining, v, xiii, 1, 10, 12, 13, 35, 50, 51, 74, 132, 151, 157, 159–162
Text to Knowledge (T2K), vi, xi, 69, 71–76, 82–84, 90, 139–141, 152
Text2Onto, 45
Thesaurus, vi, ix, x, xiii, xv-xxi, 5–10, 15, 20, 21, 23–27, 29, 31–33, 34, 39, 40–45, 48–51, 54, 56–59, 61, 62, 72, 74, 75, 78,-81, 83–86, 88, 90, 91–97, 99–105, 107, 108, 110, 112–118, 121, 123–125, 128–130, 132, 13–136, 138–145, 150–153, 155, 157, 158, 160–162
Thesaurus representativeness, 6
Tokenization, 6, 53, 74
TopicalPageRank, TPR, 51, 79
TopicRank, 51–53, 79, 82, 140, 147
Twitter, 132, 134, 145, 155, 156
Type Token Ratio (TTR), xix, 69, 137–140

U
UIMA Token Regex, 75
Unithood, 14, 15, 74, 75
Unstructured Information Management Architecture (UIMA), 75
Urlich's Web, 63

W
W3C Consortium, 10, 47, 48, 50, 148
Web Semantics, 46
White Book of Cybersecurity, ix, xv, 27, 32, 85, 92, 112, 138

Word embedding, vi, xix, xxi, 27, 41,
        51, 54–56, 94, 99, 100,
        102–104, 107, 110, 115, 117,
        142, 144, 148, 149, 151, 156,
        162
Word2vec, xi, 51, 56, 94, 99, 142, 162
WordNet, 13, 112, 57, 58

Wordsmith, 70

Y
YATEA, 18

Z
ZeroUno magazine, 63